Give 'Way to the Right

Serving with the A. E. F. in France during the World War

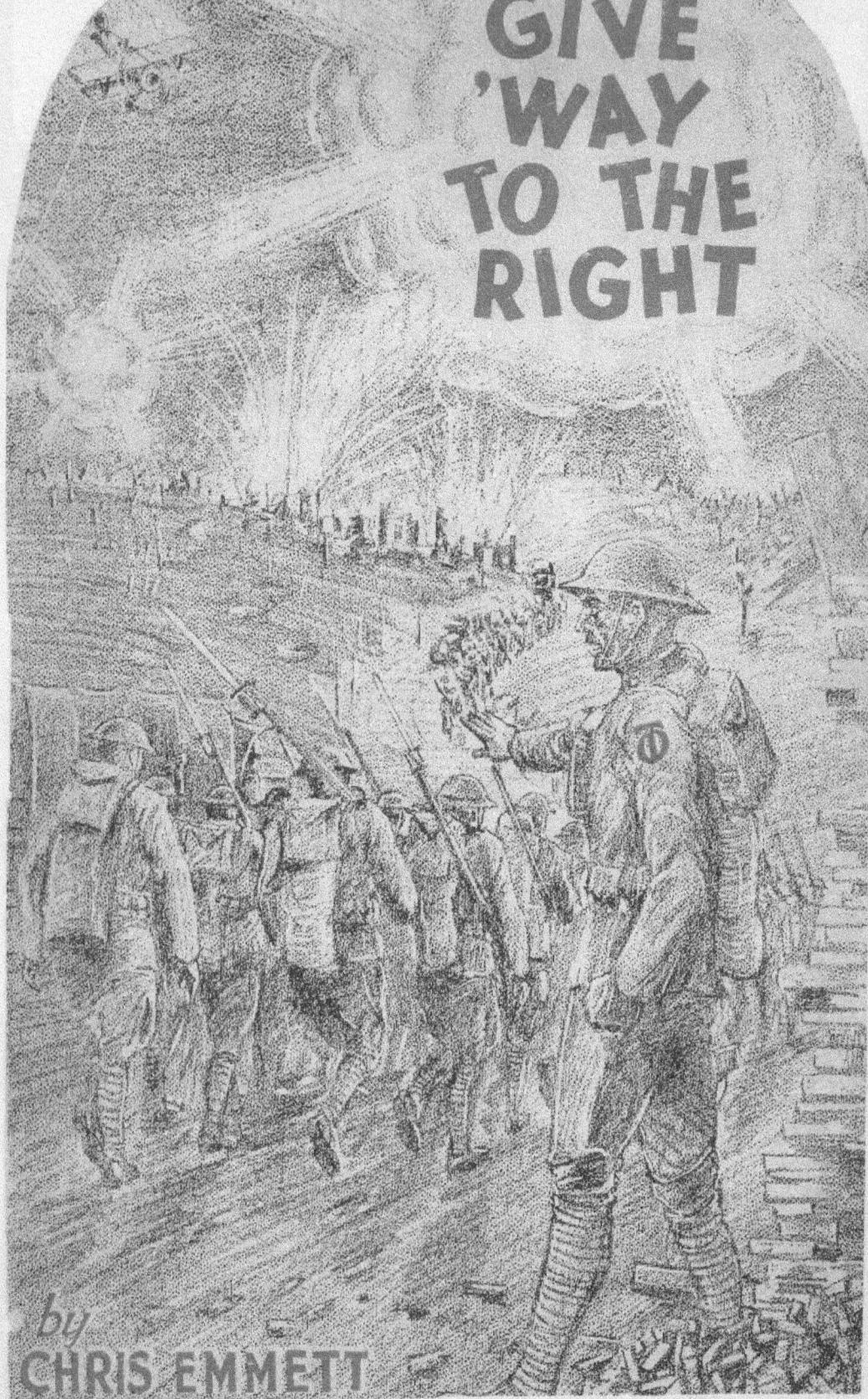

GIVE 'WAY TO THE RIGHT

Give 'Way to the Right

Serving with the A. E. F. in France during the World War

Chris Emmett

Edited by David Scott Stieghan

Blue Ridge | Cumming | Dahlonega | Gainesville | Oconee

Published by:
University of North Georgia Press
Dahlonega, Georgia

Printing Support by:
Lightning Source Inc.
La Vergne, Tennessee

Cover image: *The Menin Road*, Paul Nash, 1919
Frontispiece: Remnant of original dust jacket for *Give 'Way to the Right.*
Book design by Corey Parson.

ISBN: 978-1-940771-44-1

Printed in the United States of America
For more information, please visit: http://ung.edu/university-press
Or e-mail: ungpress@ung.edu

Note from the Publisher: This edition includes the spelling and grammatical errors that appeared in the original manuscript by Chris Emmett.

Contents

The Author

Just after discharge from the army.

THE MEN OF THE NINETIETH DIVISION
30,000 STRONG; TO
ITS DEAD, ITS LIV-
ING. A DIVISION
WHICH CONTRIBUT-
ED 8010 BLOOD
SACRIFICES UPON
THE ALTAR OF THE
WORLD-HOLO-
CAUST. TO THE
HOUSES. MILLERS,
MOLLYS, MEEKS,
TEMPLETONS,
STERLINGS, LEURS,
O'HARAS, GARNERS,
WHEELERS, TEBBES,
WHITAKERS, KINGS,
FLOREZ, ALLENS,
SINGLETONS,
GOODSONS; THOSE
THOUSANDS OF
GENIAL, PERSE-
VERING, SHRINKING,
BLATANT, COM-
PLAINING, SWEAR-
ING, GOD-FEARING
SOULS. ALL OF
THEM. TO THOSE
WHO WENT BUT DID
NOT RETURN; TO
THOSE NOW NUR-
TURING MEMORIES
OF YESTERDAY; TO
EACH OF YOU, I
DEDICATE THIS
HUMBLE NARRA-
TIVE HOPING YOU
WILL GLEAN A
LAUGH TO OFFSET
A PRESSING TEAR.

Editor's Acknowledgments

Since first reading a copy of Christopher Emmett's *Give 'Way to the Right*, it has been a passion of the editor's to make this interesting and informative work more available. A small number of the first edition were published in 1934 and the originals are getting more difficult to locate. As part of the Doughboy Series of the University of North Georgia Press, it is the goal to edit this scarce book to make it more accessible to modern readers. The original footnotes were prepared by Emmett, but those that begin with, "(Ed.)," are added by the editor. Additional illustrations, a chart, and a map are added to provide additional perspective. Nothing is removed from this edition of the original and it includes all the original illustrations and a roster of the members of Emmett's Infantry company.

The editor would like to thank the staff and faculty readers at the University of North Georgia for their professional work in improving and presenting this book. My father, Don L. Stieghan, lent his considerable research expertise in locating many of Emmett's personal and military records. To my numerous colleagues who discussed the fighting skills and tactics of the Doughboys over the years, I thank you for your insight. The editor takes responsibility for all errors and welcomes criticism. And most of all, I thank my wife, Chantay Bedford Stieghan, for her considerable editing skills as she helped make a military history book more readable.

David Scott Stieghan

Foreword

"An efficient preparation for war can alone secure peace."

JOHN ADAMS

When this narrative of events was first written it was not intended that it should be given to the public. I wrote these recollections the week following my discharge from the United States army, after having served almost a year with L Company, 359th Infantry, 90th Division.

Since then the tragic experiences have been told, and the generals have re-explained their explanations, but too few Hervey Allens, Riley Stricklands, Laurence Stallings, and Leslie Langilles have told the story of the humble soldier who executed the orders, blindly of course,—not knowing their significance,—but effectively just the same.

My experiences, I now know, were only the common ones of men of lowly rank. They represent, perhaps, a cross-section of the deeds, acts, and thoughts of the trudging doughboy, of all the men of 'that man's army'. To ex-service men they need no re-telling, but a generation that knows nothing of these things has grown up since then. In concluding to release this narrative it is my hope to convey a true picture of army-life to those who follow 'Old L Company', and show to youthful war-enthusiasts that "Its cost takes away its taste".

CHRIS EMMETT.

San Antonio, Texas.

"Some books are lies frae end to end,
And some great lies were never penn'd:
Ev'n Ministers, they hae been kenn'd,
In holy rapture,
A rousing whid, at times, to vend,
And nail 't wi' Scripture.
But this that I am gaun to tell,
Which lately on a night befell,
Is jus as true's the Deil's in hell
Or Dublin city . . ."

(Robert Bums)

Editor's Introduction

Doughboy of the Great War and lawyer, Christopher Emmett, was born on June 30, 1886 in the small town of Energy, Texas, located sixteen miles southeast of Comanche. His father, Thomas Addis Emmett, Sr., was born in Ireland about 1850, and emigrated to the United States where he married Laura Frances Pickett and raised eleven children. Education was important to the Emmetts. By 1900, during a time when many parents kept their children at home to work on the farm, four of the then eight Emmett children attended school while only one other child living nearby did so.[1]

As a young adult, Chris Emmett entered studies at the University of Texas at Austin, and then went on to pursue his law degree. While at the University of Texas, Chris became a member of the Ancient and Honorable Order of Rusty Cusses as a Reuben ("Rube") member of an anti-fraternity fraternal organization that was founded mostly for fun in B Hall. Many of the members became Senators, millionaires, and quite a few became lawyers. After passing the state bar exam, Chris moved back home with his parents who had moved eighteen miles away from Energy to Hamilton, Texas, where he began his practice of law while his father sold windmills.[2]

In 1912, Chris Emmett took a position with the Southern Pacific Rail Road in the Claims Department as legal counsel. For the next six years, he traveled across Texas appearing in court on behalf of his railroad employer. But when the United States Congress formally declared war on

1 (Ed.) 1900 United States Census for Christopher Emmett. Comanche County, Texas.

2 (Ed.) 1910 United States Census for Christopher Emmett. Hamilton City, Hamilton County, Texas. Richard A. Holland.

Germany on April 6, 1917, everything changed. The Selective Service Act, authorizing the President to increase military manpower, was passed on May 18. Like so many others, Emmett appeared for his draft registration and physical on June 5. With his only claim for exemption being a sister at home with a hearing deficiency, he was declared fit and eligible to serve in the U.S. Army and sent home to await his call to the colors from his local Selective Service Board.[3]

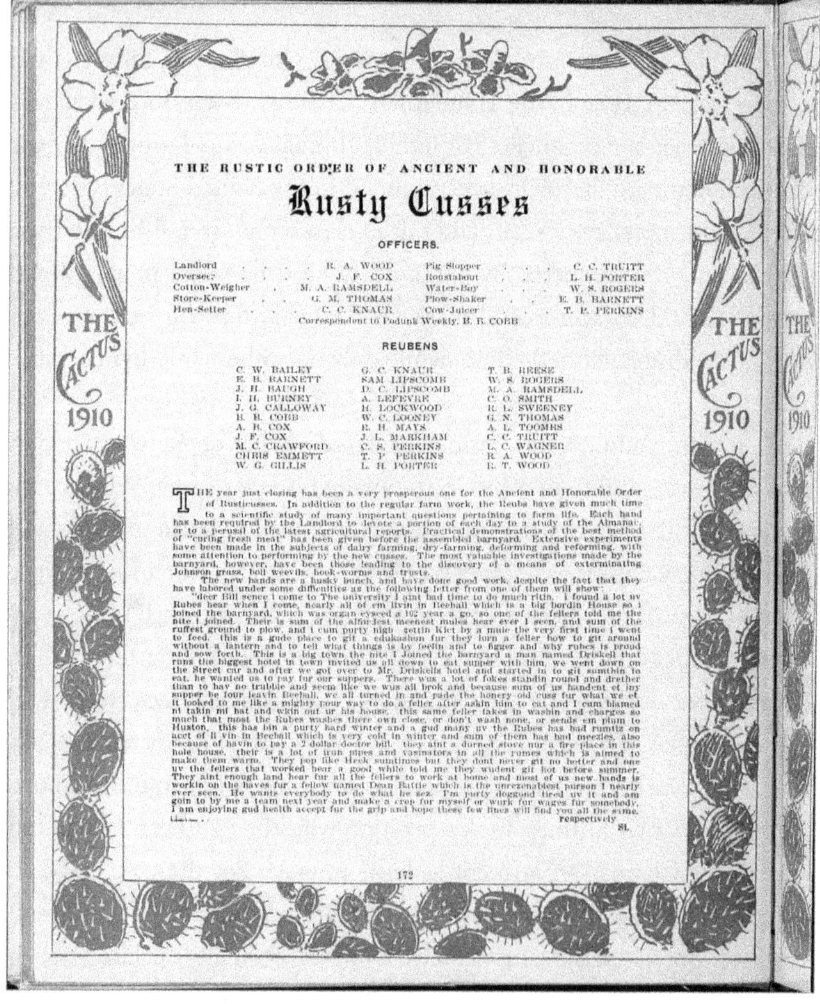

THE CACTUS 1910

THE RUSTIC ORDER OF ANCIENT AND HONORABLE

Rusty Cusses

OFFICERS.

Landlord	R. A. WOOD	Pig Slopper	C. C. TRUITT
Overseer	J. F. COX	Roustabout	L. H. PORTER
Cotton-Weigher	M. A. RAMSDELL	Water-Boy	W. S. ROGERS
Store-Keeper	G. M. THOMAS	Plow-Shaker	E. B. BARNETT
Hen-Setter	C. C. KNAUR	Cow-Juicer	T. P. PERKINS

Correspondent to Podunk Weekly, B. B. COBB

REUBENS

C. W. BAILEY	G. C. KNAUR	T. B. REESE
E. B. BARNETT	SAM LIPSCOMB	W. S. ROGERS
J. H. BAUGH	D. C. LIPSCOMB	M. A. RAMSDELL
I. H. BURNEY	A. LEFEVRE	C. O. SMITH
J. G. CALLOWAY	H. LOCKWOOD	R. L. SWEENEY
B. B. COBB	W. C. LOONEY	G. N. THOMAS
A. B. COX	R. H. MAYS	A. L. TOOMBS
J. F. COX	J. L. MARKHAM	C. C. TRUITT
M. C. CRAWFORD	C. S. PERKINS	L. C. WAGNER
CHRIS EMMETT	T. P. PERKINS	R. A. WOOD
W. G. GILLIS	L. H. PORTER	R. T. WOOD

THE year just closing has been a very prosperous one for the Ancient and Honorable Order of Rusticusses. In addition to the regular farm work, the Reubs have given much time to a scientific study of many important questions pertaining to farm life. Each hand has been required by the Landlord to devote a portion of each day to a study of the Almanac, or to a perusal of the latest agricultural reports. Practical demonstrations of the best method of "curing fresh meat" has been given before the assembled barnyard. Extensive experiments have been made in the subjects of dairy farming, dry-farming, deforming and reforming, with some attention to performing by the new cusses. The most valuable investigations made by the barnyard, however, have been those leading to the discovery of a means of exterminating Johnson grass, boll weevils, hook-worms and trusts.

The new hands are a husky bunch, and have done good work, despite the fact that they have labored under some difficulties as the following letter from one of them will show:

"deer Bill sence i come to The university I aint had time to do much ritin. i found a lot uv Rubes hear when I come, nearly all of em livin in Beeball which is a big bordin House so i joined the barnyard, which was organ-eysced a 102 year a go, so one of the fellers told me the nite i joined. Their is sum of the alfirdest meenest mules hear ever I seen, and sum of the ruffest ground to plow, and i cum purty nigh gettin Kict by a mule the very first time i went to feed. this is a gude place to git a edukashun fur they lorn a feller how to git around without a lantern and to tell what things is by feelin and to figger and why rubes is proud and sow forth. This is a big town the nite I Joined the barnyard a man named Driskell that runs the biggest hotel in town invited us all down to eat supper with him, we went down on the Street car and after we got over to Mr. Driskells hotel and started in to git sumthin to eat, he wanted us to pay for our suppers. There wus a lot of fokes standin round and drether than to hav no trubble and seem like we wus all brok and because sum of us handent et iny supper be four leavin Beeball, we all turned in and pade the honery old cuss fur what we et. it looked to me like a mighty pour way to do a feller after askin him to eat and I cum blamed ni takin mi hat and wkin out ur his house. this same feller takes in washin and charges so much that most the Rubes washes there own close, or don't wash none, or sends em plum to Huston. this has bin a purty hard winter and a gud many uv the Rubes has had rumitiz on acct of li vin in Beeball which is very cold in winter and sum of them has had meezles, also because of havin to pay a 2 dollar doctor bill. they aint a durned stove nur a fire place in this hole house. their is a lot of irun pipes and vasinators in all the rumes which is aimed to make them warm. They pop like Heck sumtimes but they dont never git no hotter and one uv the fellers that worked hear a good while told me they wudent git hot before summer. They aint enough land hear fur all the fellers to work at home and most of us new hands is workin on the haves fur a fellow named Dean Battle which is the unrezenablest purson I nearly ever seen. He wants everybody to do what he sez. I'm purty doggond tired uv it and am goin to by me a team next year and make a crop for myself or wurk fur wages fur somebody. I am enjoying gud health accept fur the grip and hope these few lines will find you all the same.

respectively

SL.

172

3 (Ed.) "Medium height, Slender build, Gray eyes, Brown hair, Bald- no." Registration Card. Form 1. 1815/13. Registrar's Report, 42-2-6-A. Precinct 15, Cherokee, Texas. *Texas History Online*, accessed March 3, 2014. https://tshaonline.org/handbook/online/articles/fem02; https://www.archives.gov/research/military/ww1/draft-registration

Emmett was an accidental Doughboy. A bachelor lawyer who barely missed the draft age cutoff of thirty-two by three weeks, Emmett was willing and eager to serve. Seeking adventure and responsibility, he was excited by the prospect of selection for officer training. After all, he was smart and a successful lawyer in civilian life and figured he would naturally fit right in as an officer. After an unsuccessful attempt to enlist in the Air Service to become a pilot, Emmett appeared in France as a private with a rifle on his shoulder about to embark on a new quest as a Soldier. Although it was not the role he wanted or expected, he came to relish his experience as an enlisted man.

Christopher Emmett was a distinguished, intelligent, and cultured man who did a rough thing and enjoyed it, unexpectedly. He was not a good soldier because he was always breaking the rules. He often snuck away to enjoy life and experience culture when no one was watching him, but he was a good fighter. At first, he felt that he had risen from his station as a farmer's son to become a lawyer only to be relegated to the lowest rank in the military. He began the war as a patriot fighting for his country, but soon came to the realization that he was actually fighting for the men around him and the comradery they shared. Dispite his seeming lack of discipline, Emmett's commmanders saw that he had leadership qualities and began to assign him important roles as they neared combat. When his leadership was needed, he stepped up and did what was necessary.

Give 'Way to the Right is the unvarnished tale of an uncommon man who served as a common soldier in the First World War. Fifteen years after the Armistice, and the high-profile officers had written their accounts, Emmett decided it was time to tell his story. Not only did he want to explain the Doughboy experience to a new generation of youth raised after the war, but he wanted to inform his contemoraries who did not serve what it was really like Over There. The fact that few seemed to understand or appreciate the role of the common soldier is what motivated him most to write his book. He was proud to be a soldier and wanted the world to know the simple infantryman fought the bulk of the war. Many of his friends died fighting and he did not want them to be forgotten. As they marched toward the Front, they gave 'way to the right for all others to pass. Emmett could not neglect his chance to tell the world their story. He decided to add his version of the Doughboy experience to the postwar chronicles of other Soldiers.

Arthur Guy Empey had published his best-seller *Over the Top* in 1917 and a number of other veterans had put their experiences down on paper within a few years following the Armistice. At first, they were very proud of their service in the Great War. They, with the support of the entire nation, saved from defeat the nations of the Western Allies making the world safe for democracy. The literature of the Doughboys was for a few years that of youth excited to have gone and participated in the greatest event of their lives. Each of the combat divisions of the American Expeditionary Force prepared their unit histories within a year of the Armistice while the staffs were still together on duty occupying Germany or demobilizing together at camps in the United States. The last chapters of each history was filled with the honors of the division heroes, as well as a memorial list of all those who died or were killed in action. They felt strongly about remembering their accomplishments, and those who fell, before they took home a unit history that celebrated their part in the victory. After their last parades upon returning home they pinned the crimson discharge stripe on their coat sleeves and basked in the honor of their community and family before they took off their uniforms and returned to their lives as civilians.

Within a few years, disappointment over the results of the peace took its toll on their pride of accomplishment. A malaise took hold among many of those who had sacrificed a year or two of their youth and lost comrades in the struggle. They began to write of their experiences as participation in a waste of treasure and lives. Some considered themselves eyewitnesses to the slaughter of the Lost Generation and wrote of their participation as innocence lost and as observers to a tragedy. By the middle Thirties, the boom and bust times of the Roaring Twenties and the Great Depression focused attention away from the war. It caused many former Doughboys to watch with disgust the rise of Communism and Fascism in many of the very countries in which they fought to halt imperialism in their idealistic youth. As they matured and approached a reflective age, the world went to war with itself again. Many watched as their sons donned their country's uniform to go and serve again in another world war.

While Emmett did not write *Give 'Way to the Right* as a heroic tome or an expose', he did have examples of why he should write and how. He mentions by name in his foreword four authors of other Doughboy memories whom he admired and who inspired him to add his tale to the shelf of memoirs. William Hervey Allen, Jr., wrote the acclaimed, *Towards the Flame*, a chronicle of his intense combat as an Infantry company commander in the 28th Division. Allen became a well-known poet and also wrote a biography of Edgar Allen Poe and the best-seller-turned-into-Academy-Award-winning-movie, *Anthony Adverse.*[4]

Another East Texas Doughboy from Longview who also served in the 90th Division, Riley Strickland published his lively memories of serving as a combat Engineer in 1920 as, *Adventures of the A.E.F.* It is possible that he and Emmett knew of each other and perhaps rode on the same train to Camp Travis, Texas, in 1918.[5]

One of the most successful writers of the former Doughboys was a former Marine lieutenant from Macon, Georgia, Laurence Stallings. His autobiographical novel, *Plumes*, is closely based upon Stallings' own Great War combat that resulted in the award of the Silver Star and the eventual loss of both his legs. He co-wrote the hit play and film screenplay, *The Big Parade* (1925), penned the largest grossing silent film screenplay in history, *What Price Glory* (1924), and the screenplays for three other important films brought to the silver screen by John Ford, *Three Godfathers* (1948), *She Wore a Yellow Ribbon* (1949), and, *The Sun Shines Bright* (1954). His two photographic treatments of the World War and his anthology of the common American warriors, *The Doughboys: The Story of the AEF, 1917-1918* (1963), placed him among the most recognizable American writers on the war.[6]

Leslie Langilles published his memoirs one year before Emmett. In, *Men of the Rainbow*, the former corporal of Battery B, 149th Field Artillery

4 (Ed.) Hervey Allen, *Toward the Flame.* George H. Doran Co. (New York, 1926).

5 (Ed.) Riley Strickland, *Adventures of the A. E. F. Soldier.* Von Boeckmann-Jones Co. (Austin, Texas, 1920).

6 (Ed.) Laurence Stallings, *Plumes,* (New York: Harcourt, Brace and Company, 1924).

Regiment, 42nd "Rainbow" Division, also included a great deal about Soldiering as a cannoneer at the front. Though these four authors were perhaps not the best known of the Doughboy chronicles, Emmett was inspired by their prose and their stories to add his version of events in his own way.[7]

7 (Ed.) Leslie Langille, *Men of the Rainbow, by Leslie Langille of Battery B, 149th Field Artillery, U. S. A. 42nd (Rainbow) Division, A. E. F., (Chicago, Illinois: The O'Sullivan Publishing House, 1934).*

Chapter 1

Fifteen Years Afterward

"O, the infantree, the infantree,
With dirt behind your ears,
The cavalree, artilleree,
And the — — engineers,
But they'll never rest the infantree
In a hundred thousand years."

Bands played. Discordant voices sang. Thousands of ill-fitting and faded uniforms turned the city streets to a musty olive-drab. Throngs of convivialists chanted words more suggestive than expressive. A genial comradeship pervaded the throng. Older men without uniforms stopped and listened. A few patted their feet in rhythmic harmony. Boyish-men shoved and pushed their way through the congestion. A man of some forty years stepped gingerly back from the surging passersby. His face was softer than those of the tramping, chanting crowd. His expression was inane. There was a meticulous nicety about his apparel. He looked about as if to find someone with whom to speak, perhaps to express his disgust, as a tramping group of improvising chanters crunched past his toes. "I have never been able to understand," he spoke, "why these men, (now that the war has been over fifteen years) cannot forget their vile songs and conduct themselves with decorum,—if they must have these conventions!" . . "And you were 'Over There' with them?" I inquired sarcastically. "You seem to be of their age." . . . "No . . . eh . . . but I performed my duties, as any good man must, with the Y.M.C.A. throughout the entire period of the war." . . . But I was moving on. I must not answer. One look at his face showed his fatuity . . . And with clenched teeth I mused to myself. He could not understand. Ability was lacking . . . And there were those who had profited by Mar's unleashing, also, who did not want to understand. But there were some who did understand . . . and how indulgent and kind they have been all these years! . . . But youth does not understand. Perhaps these boys have not been told in terms true to the glories, to the miseries, of those days? And if they were told would they have an understanding of the value of peace? Would such an understanding make them more inflexible in the surrender of the virtues of their government? Could they understand with what these old military friendships have been congealed, and could they not be made to know that, even though war has its glories, "Its cost takes away its taste." (*Le couet en ote le gout.*)

Chapter 2

Engulfed by War Clouds

In the early days of 1917, as I sat in a restaurant in Tyler, Texas, reading the news of the World War, Peyton Lane, with whom I dined, called my attention to the fact that, according to the newly enacted Selective Service Act, all men must register for military duty who had not attained the age of thirty-one years and who were over twenty-one. Lane grimacingly advised he would be thirty-one years and one day old on June 5, 1917. I lacked twenty-five days of being over the prescribed age. This difference of twenty-six days in our ages caused both of us no little merriment,—with the honors to go to Lane.

Even when I look back at the situation at this time in retrospection, when all has been done, I see it was a disquieting piece of news I had received. I think I have seen amongst the American people very few moral or physical cowards, but all of us, even at that time, had read enough of the horrors of the war to have an unconscious dread of all things connected with the army. I see now that it was so little I actually knew of its realities. Had I and all others, then, known as much about the army as my generation now knows the American Expeditionary Forces would have been inducted lacking that bouyant spirit which was one of the most potent contributing factors to the triumphant victory over the enemy forces.

On registration day I drew No. 13 and made plans to enter the army, intending to be transferred to the Intelligence Department, the military branch of the Department of Justice. First, I attempted to volunteer into that department, but found after I had made a trip to San Antonio that I would not be accepted because of a ruling that all men of registration age must be inducted through the selective service process. The rule provided, however, that after induction one might be transferred upon application.

While attempting to enlist at San Antonio, I became acquainted with a Major Barnes of Fort Sam Houston, who discussed with me at length my education, previous training, etc., and promised that my request for transfer would have his approval when I arrived at camp with the drafted men. This promise Major Barnes faithfully kept. It was not known to me then,—but there is a possibility that it was known to the major,—nor did I learn of this until after the close of the war,—that the road to a commission had been paved for me by a friend. This arrangement, however, was frustrated, as shall be disclosed in its proper chronological order.

After leaving Major Barnes I witnessed a military parade through the streets of San Antonio. Then I realized my "war-temperature" was running rather high. The stream of olive-drab, the legato tread of thumping feet upon bricks, and finally the jaunty drum-major leading his rhythmically divine Cecelias before the plaudits of the cheering throng, sent my pulsating blood out of control.

"Bugles that whinnied, flageolets that crooned,
And strings that whined and grunted"

had captured the reason of the crowds, and under the spell of music and mob-psychology I rose to ecstatic heights feeling that "Even so shalt thou live in the echoes of Fame." But there was one who stood apart, as the echoes died away in the distance, to remark: "Were all bandmasters killed at birth, there would be no wars." My fervor, however, was not to be cooled by his observation,—as wise as I came to believe it was,—and I rushed off to Kelly Field to join the aviators. Here again I was advised, when I disclosed my age, that I must abide my time. But the officer, as I turned away added rather wryly, "Besides you are too old to be indiscreet." It was months later before I understood what he meant to imply: "The younger the boy, the more venturesome; consequently the better aviator."

I was later examined at Rusk by the Local Board and was found physically fit. I was then told that June or July would be as early as I would be called to report for military duty. Relying on this, I made my plans accordingly, but despite this statement (and I now see it was a statement which no Local Board member had a right to make, for the exigencies of war did not warrant any promises) upon my return from Shreveport, La., early one morning I was met in Jacksonville, Texas, by Judge Early[1], who gave me a shock.

The Judge was a man of towering stature and solemn demeanor. He had a habit which was amusing to observers. He spoke very deliberately, and interspersed in his sentences he would pucker up his lips and whistle a little before proceeding. The Judge accosted me with "Well . . . whew . . . whew . . . whew ... I guess I ought to tell you . . . whew-uh . . . Your name is on the board." After being thus in sibilant tones informed of the impending developments I went to the Post Office where I found, in fact, my 'name was on the board' . . . and . . . "whew-whew" . . . I had only ten days in which to attend to my personal affairs and report at Rusk for military duty.

The ten days fled as I went to Houston to be relieved of my employment in the Claim Department of the Southern Pacific Railway Lines, then to Hamilton where I visited my home-folks, and back to Jacksonville, only to ride to Rusk, on April 26, with Ben Whatley, where we, along with thirty others, were sworn into the service of the United States, to be "Soldiers of the United States of America," according to our "Greeting."

Through the kindness of Bill Simmons, Chairman of the Local Draft Board, the men from Jacksonville were permitted to return to Jacksonville to remain until the International & Great Northern train came that night upon which we were to entrain for Camp Travis. I was 'appellated' sergeant of the detachment of rookies being, of course, responsible for their appearance at the train. I accepted the responsibility lightly thinking no man would take leave according to the French-manner. I had, however, some difficulty that night making convincing explanations to Simmons why Ben Whatley had disappeared so rapidly after I had called the roll shortly before the departure of the train. The facts were, of course, Ben had never responded to his name but had taken pleasure in answering the call

1 Judge Early, Mayor of Jacksonville, Texas.

of a lady-friend relying upon me to skip his name which I 'inadvertently' did. Ben's timely arrival, however, re-established Simmon's faith in my avowals, and only Ben, the girl, and I knew.

The day of the 26th of April was spent with our friends. Women of the Red Cross served a bounteous departing repast, and when the train coupled into our special car, we sank down upon cushions, tired and happy. The train rolled out toward San Antonio and little did we realize what the next eleven months had in store for us! The thing which impressed me most that night was a talk I had with Rogers Tipton. He was very morose. He would have it that he had seen his last day as a civilian, avowing most solemnly that he would not live to see the end of the war.

Our sleep that night was nil, of course. Rogers Tipton, Joe Templeton, Wm. L. (Bill)) Goodson, Jim Alexander, Ben Whatley, and a few other bacchanal sports of the 'silk-stocking' brigade spent the night in crap-shooting and other conviviality, re-invigorating ourselves at Taylor where the train stopped a short time the following morning. Here we were 'issued' our first army breakfast which consisted of an apple, sandwich, and a cup of lukewarm coffee. With these and our night of dissipation behind us we became converts to the tenets of the Food Administration.

Arriving in San Antonio we were shuttled off to Camp Travis. We detrained at 1:30 P. M., April 27th, and were not long in learning the true value of liberty. Our every movement was supervised by a soldier who was acting, of course, under orders. Even our execretory manifestations were directed, if not supervised, and I harked back to my lineal Irish ancestor and wanted to fight, to fight for liberty and democracy, especially liberty. They 'herded' us through the 'Bull-pen'[2],—a constant stream of flowing, cursing, protesting, complaining humanity,—White-men, Mexicans, Negroes, all together,—so that the all-wise medicos,—(soon to be the much detested medicos,—accused, and unjustly, of seeking an 'easy-berth,') could pass judgment upon the physical fitness of each man. It was strange to me, then, (although it is not so now) that I did not see a single man passing through the 'Bull-pen' disqualified by the physical examination. I see, now,

2 The Bull-pen was so called by the men because all incoming soldiers passed under the observation of the officers like a drove of cattle.

it could not have been otherwise, for to have shown to any of the men then entering the army that duty might be avoided by physical incapacity would have magnified in the minds of many men their minor deficiencies and resulted in greater difficulties in classification and induction of recruits. No doubt there were many physical deficiencies noted at that time and reclassification and discharge effected later on.

Before arriving at the 'Bull-pen', walking en route from the train, I met an early boyhood chum, Howard Edmiston,[3] whom I had not seen for many years. He came to me and made himself known. I was to see him twice after this, once at Camp Bullis and again in Winneldown and Winchester, England, where we had a most unusual day.

After a prolonged stand in the sun to which we were then not accustomed it became apparent *probably* to the officers in charge that we had not eaten. We were, therefore, driven in 'herd-formation' (for none then even knew line-formation) to a barrack where a very *delectable* meal had been spread for us. It consisted of *cold beef and bread, without water*. We were issued mess-kits from which we were thereafter to eat, and Ike Lawler, sensing the "coming of the future" opined to me: "Well, Chris, we have hubbed hell." I agreed with Ike, even then, but the hell we had hubbed there was the attar of roses compared with the hitch most of us later did in Europe.

The 'medicos' finished "marking and branding" us at the Bull-pen about 9:30 p. m., and off we went to 30th Company, 8th. Battalion, 165 Depot Brigade. Some were separated from the main contingent, and from that time on we were to become more and more scattered until few were left together who had sworn allegiance to our government at Rusk on that eventful day.

At 30th company we were given a blanket and a mess-kit of our own, which two articles were to be our only earthly comforts for days to come. To those of us who had been reposing on the downy beds of ease pillowed with the comforts of life's luxuries, a spring-cot, (the wires of which cut the map of a disturbed Europe across my fragile anatomy,) and one blanket did not comport with my idea of living! But my ideas of what were the

3 Howard Edmiston was later to be one of the casuals of war, having been disabled back to his home at Hamilton. Texas, by a gunshot wound.

necessities of life were soon to undergo a vast and devastating change. What was once considered an imperative necessity changed to a luxury unattainable. Men learned to do without and profited by the sacrifice, becoming more and more hardy and self-reliant as they discarded desires for physical aids to comfort.

Needless to say we soon fell asleep that first night as soon as the opportunity came, but this sleep was not to be long until we were awakened by that most disliked of all bugle calls, 'reveille',—the first we had heard,—but it was to be with us evermore! Following immediately after the 'first call' came a shrill whistle and a command to "Fall Out!" Some self-opinionated wit, pushing his head out an up-stairs window, wanted to know "where we were to fall to." A sergeant answering in raucous tones, embellished with unsavory profanity, quieted the wit and we "fell out for reveille."

Reveille, to the soldier, is the most hated of ceremonies. Were a man to live in the army until the government on its own initiative concluded to pension him, no soldier would learn to like its sounding. This, I believe, particularly applied to me for my nocturnal habits had taught me the uselessness of early rising!

The day following our advent into Camp Travis was not an eventful one. It was Sunday. Rookies rankled in flesh when it became known the area was quarantined and all of us were proscribed from visiting the city as the older men in the service were doing. Bill Goodson made friends with a Mexican corporal who was the happy possessor of a pass which permitted him to come and go from the camp at will, and Bill, fast learning the ethics of the army, was answering to the name of "Sanchez" and disporting himself on the streets of San Antonio, exhibiting this 'hombre's' pass as the evidence of his right. How Rogers Tipton passed the guard he never told us but he returned to quarters carrying spiritual evidence of his visit to the city.

George Singleton and 'Doby' Lloyd who braved the early days of the formation of the 90th. Division,—Cherokee County friends of the new contingent,—visited us, primarily to "wise-us-up" to the ways of the army and prepare us for some of the trials to come. Doby's tales of the army were even then most 'worthy' ones, but they were the acme of simplicity compared with his effervescings after the war. And Doby went places and saw things, too!

Days followed and we were 'fitted out' in khaki, discarding our civilian clothes. I have used the word 'fitted' but this, truly, is an inaccuracy. A sufficient number of articles of clothing was issued to each man, but if the clothes did not fit him (and they never did), it was not considered the fault of the government. It was up to the soldier to trade or exchange his clothing with other soldiers until he had a uniform which he could wear. Being of a 'trading disposition' I was soon bedecked in a uniform the like of which has covered the anatomy of few men. I had gone about my trading in the reverse order,—that is, I traded for every garment which would not fit me, and in this uniform I 'stood formation.' The captain of the company spied me, whom he could not very well over-look, and when the formation was concluded he ordered me to remain. The following conversation ensued:

Captain: Private, where in the hell did you get those clothes?

Private: Sir: They were issued to me.

Captain: Do you realize your appearance is a disgrace to the United States army?

Private: Sir: I had not thought my appearance particularly commended the service.

Captain: Commend! Hell! Get those clothes off and don't show yourself on a company-street again with such a garb. Sergeant (Calling a quartermaster sergeant), take this man to the supply depot and have him properly dressed, . . . *and do it now!*"

I followed the sergeant into headquarters. He, too, was laughing, and gave me the choice of the supply-room. I left him presenting a much better appearance than the 'traders.'

Chapter 3

Innoculations

To fit a man properly for the service it was the belief of the medical branch of the army that all men must be inoculated with typhoid and para-typhoid serums. They added to this, for good measure, a vaccination against smallpox. I am not sure these were as indispensable to health as they would have us believe, but these precautions unquestionably assisted in allaying many diseases. I am equally as sure that the inoculations and vaccinations caused much real pain.

Soon after our arrival at Camp Travis we were again 'herded up in mob-formation' and conducted to the infirmary where a most careful physical examination was made of each man. At this time one "shot," as the boys called it, was given and at the same time we were vaccinated against smallpox.

The manner of giving the smallpox vaccine was not attended with any technique. The vaccine came prepared in a glass tube and, as the soldiers passed in line, the tubes were broken and each man was cut across the arm with the jagged glass, inflicting a ragged wound. If the soldier flinched or cried-out then he was slashed again for good measure.

Each week after our first appearance we were returned to the infirmary until three shots had been given. I never had any of the diseases which we were supposed not to have, but if the diseases were worse than the preventatives, I am certain I escaped death in its most violent forms.

My good friend, Joe Templeton, had gone to the army disporting a waist-line of forty-two inches. Each 'shot' reduced him in corpulency so rapidly that he was kept trading clothes all the time to prevent appearing like the forsaken vagabond. I did not have sufficient avoirdupois upon entering the army to be of any importance, weighing only 119 pounds, but even some of this weight was lost during the painful inoculation period. It was a common sight to see a man walking along a company-street and without warning topple over on the ground. Joe Templeton and I had gone early one morning to a restaurant, or canteen, near the barracks, and he began to zig-zag like a man over-fed on the vin-rouge of the distant land. Then he fell into the arms of a passing officer who must have thought he was rescuing a dying rookie. I came to his assistance and we administered cold water to the pate of his hairless head, and as he protested against going to the hospital, I got him back to his bunk where he spent considerable time in recuperation.

One afternoon at 'retreat,' Rogers Tipton fell prone on the ground as the last echoes of the band died away over the hills. I happened to be standing directly in his rear and caught at him but without avail, for he fell forward and skinned his face on the pebbly ground. Some officers were humane enough to permit these embryonic soldiers to rest after each 'shot,' but, of course, there is always in-humanity amongst men and many suffered unreasonable misery because of being forced on duty while the effects of the inoculations were still present.

Williard Newton, a fine fellow, became so ill from the inoculations that he missed the "tub" and was unable to go over-seas with the 90th Division. As a *just punishment* for his *weakness* he was retained in America and commissioned lieutenant! No provision for after-examination was made for the men who had been inoculated and vaccinated. I saw one man become temporarily deranged from the effects of the injections, and another in a fit of irresponsibility cut his throat. I saw another refused medical attention, when his condition was really becoming desperate, with the curt dismissal: "No use to be calling on the medical department just for a sore arm." And I saw that boy later. His arm had become so badly infected from lack of proper dressing that a huge hole sloughed off and the pain was unbearable. When he did receive attention it was feared he would lose his arm.

Chapter 4

Learning to Drill

While with 30th company we learned very little about the army. We were given 'close-order' drill, schooled in the manual of arms, etc., but the chief thing which the official family seemed to desire we should know was: the proper manner for us to comport ourselves around officers. Of course, we were advised that officers were not considered the superior of men. This, in many instances, all of us knew to be a fact, but many officers, especially those new in the army themselves, and called by the common soldier the "ninety day wonders," had not themselves learned that fact. They so conducted themselves, so the new soldiers thought at least, as to leave the impression they were superior creatures. Out of this lack of understanding between man and officer came the common-soldier's renowned disrespect for army officers. During these days it was a common thing for a soldier to recognize an officer, all booted and spurred and frilled with the embroidery of position, whom he had known as a civilian scarcely able to subsist in the economic competitive world, and be forced to salute this officer, and the soldier felt he had been humbled. With this spirit of resentment in the ranks it is a great wonder to me now that more soldiers did not get enmeshed in military insubordination. I am inclined now to the belief that officers were much more considerate than soldiers thought them to be. I can also understand the importance of a proper respect for army rank. The question was one, not of personalities, but of implicit obedience to command the lack of which transposes an army into a mob.

Charles F. King, of Dallas, Texas, a quick-acting, free-thinking youngster, first distinguished himself in the army by his refusal to salute an officer whom he had known as a civilian unworthy of personal recognition. When King was accosted by a sergeant who saw the intentional act, he again refused to salute and the incident wound up by King pummeling the sergeant around on the grounds to the invigorating refrain of "Retreat." I say this "wound up the incident" but it did not, for King found himself on his second day in the army beginning a "thirty-day stretch in the guard-house" for refusing to salute an officer, fighting, and failing to stand at 'attention' while "Retreat" was being sounded. King's service in America, therefore, was not very conspicuous, but before they could incarcerate him in the 'brig,' he had attracted the attention of the regiment with his one-string-guitar which he carried under his arm wherever he went, strumming it with an affection, whenever the opportunity was afforded, adding, of course, his vocal improvised versions of "this man's army" as he sauntered along. Being released from prison on the fifth week after his entry into Camp Travis, he was transferred to L-Company, 359 Infantry, and within a short time thereafter was on his way across the waters. His reputation had spread amongst the boys: he was not of the common kind. He became a favorite with the men. A friendship ripened with many a man because of his bouyant carelessness, a friendship which he deserved, and which helped many a man through days of gloom.

I think that King, like most men, had as much patriotic respect for the insignia of military position as America required of her soldiers, but I think, also, it was impossible for him to condescend to the 'inferiority stuff' too often emphasized by men of position, and he by nature resented the implications and was willing to suffer the consequences.

On the last day we were with 30th Company I had the misfortune, as did Ben Whatley, to have to go on guard. The night was rather a cool one, and of course, we slept none. A guard is supposed, when doing regular shifts, to walk post two hours and then go off duty four hours, subject to call. During these 'off-periods' the officers never failed to call you for some reason, or lack of reason. This night will stand out vividly in my mind for years.

I was on post at the ware-house where Ben's walk intersected mine.

We soon gauged our trips to meet at the crossing of our paths where we tarried for a little chat before passing on. Most children in their infancy have been threatened by a thoughtless mother and nurse with the 'booger-man' out in the darkness. Boys, in the years of their indiscretion, are kept at home through an unfair emphasis of the lurking dangers of the darkness. Men grow to soldier-age with an apprehensive irresolution for the night. A 'rooky' who walks his post for the first time reverts to a consciousness of 'night-qualms.' Noises which would not attract his ear in the day-time cause him to tread his post with muffled foot and straining ear. The silence of the night carries the breaking of a twig or the click of a gunner's trigger far out into the night. Breaking upon the ears of the uninitiated soldier, these sounds were premonitory of lurking dangers. Eyes unaccustomed to the darkness picked out aeriform figures which in the imagination of the 'rookie' floated or skulked noiselessly just out of reach. And to meet your flesh and blood comrade at the crossing of the path was an unspoken solace to each. How well, indeed, has the poet been able to make us see this Nemesis[1]:

"What soldier walking his dark post
Has not seen him pass,
Gliding through the shadowy wood,
Crawling through the grass.

And they who watch beside the dead
In a silent room,
Have they not glimpsed him at the door,
And heard him in the gloom?

He follows down unlighted streets
Or waits not far ahead.
He tries each door within the dark,
And stares behind each bed."

1 Patrick D. Moreland: Nemesis.

And when a dog did come out of the gloom that night, faithfully to walk the post with me, a hand on his furry coat and a lick of his tongue was necessary for me to believe him to be a reality, not a ghost. Arriving near midnight my canine friend remained with me until the eastern lights shoved back the canopy of night. Pacing a few steps ahead he guarded my progress to the end of my walk where he would repeat the procedure as long as I was on post. When daylight made its appearance, he walked off into the thinning darkness, looking back over his shoulder at me as he went. I saw him no more. He had done his duty. I had often heard of 'the soldier and his faithful dog' but this was my first experience. It was not, however, to be my last. Many times thereafter in France we had occasion to make use of captured dogs which became our faithful and helpful friends.

I had twenty-four hours of guard-duty, and during the day was assigned the duty of transferring prisoners. The seriousness of this assignment did not impress itself upon me until after the transfer had been made. I then realized I had unwittingly confronted a serious situation, one which might easily have culminated in trouble for me. I might even have been forced to kill an American soldier.

I was called from the guard-house and asked if I knew the location of the hospital. I said I did not, whereupon I was assigned a guide, who was instructed to return upon my being directed to the hospital. I was given a memorandum for delivery to the commanding officer. I went to the hospital and reported, and was asked if I were armed. I was. The officer then instructed me to go with a sergeant. I filed through the hospital and into a room with bars. Here a soldier-prisoner was delivered to me. I was then commanded to deliver him to the guard-house. The prisoner came with me willingly until we came abreast a canteen. There he began to talk, saying he had been badly mistreated; that he had come into "this man's army" unwillingly and was going to stay in it the same way; that he had no intention of doing any duty whatever unless forced to do it, and that he was going to "see the wide-open spaces poco tempo." I laughed at him, calling his attention to the song heard at every turn:

"You're in the army now;
You're not behind a plow;
You---of--- a---;
You'll never get rich;
You're in the army now."

I told him he had better make the best of it and go along with the fellows. His chief grouch seemed to be that he had been deprived of the use of tobacco since his arrest. He expressed a great desire to smoke but said the officers refused to let him for fear he would burn the hospital. I asked him if he would be satisfied with a "fair substitute" and turning to the canteen, bought him some chewing-tobacco. To my complete consternation he showed his education by quoting Charles Lamb: "For thy sake, Tobacco, I would do anything but die."

I concluded he had intended to make a break for liberty here and advised him of the uncertainties connected with such an effort. He was made to know I had been sent *after* him with explicit instructions to return him to the guard-house, and to the guard-house he was going even though I had to put a few holes through his contrary carcass and carry him on my shoulder. He then said: "I believe you'd do it." Thereafter his desire was not so keen to leave for parts unknown. I helped him cut the seams in his jacket concealing his much desired tobacco, then delivered him into the guard-house. I was asked, of course, when I arrived with him if he had concealed anything in his clothing. I said he had not, and much to my surprise he was not searched.

While enroute with this man he informed me that he had come to the army from Arizona; that life had been unsatisfactory with him; that he had been sent here, and reiterated he didn't intend to stay. He "represented" himself as being a "hard customer," as having been once convicted for murder and it was "all in a day" with him. I confirmed some of these statements later in a talk with the officer-of-the-day who also said he had caused a great deal of trouble in the hospital. I saw him several times during the day basking away his time in the "brig."

Upon returning to the company from guard-duty I learned our company was actually waiting for Whatley and me, and we were to move! I could not, of course, see it was the first of a series of moves which would ultimately bring us into contact with the German forces. It was late afternoon when Whatley and I rejoined the company which moved hastily across the camp where we stood on a drill-field for what seemed hours to me. It was then concluded by the 'Powers that Be' that we would not be placed in companies until the following morning, and we were rushed off to a new barrack in another part of the camp, distant from our old quarters, and with no supper we were released. I was dead-tired from twenty-four hours guard duty and was soon in bed, only to be awakened and called to the 'orderly room' where I was informed my record showed I was possessed of some education and that I was to do some stenographic work during the night. I professed a complete willingness to do any service within my abilities but disclaimed any clerical skill. I 'got by' with 'this one' and started to bed again only to be stopped by an old friend, R. R. Bondurant, who had come to advise me he had seen in Major Barnes' office a request by the Major for my transfer to his department. But as much as I desired to visit with my old chum, I fell asleep on my bunk and my next recollections were the painful blasts of "reveille" call.

We had a scant breakfast, and I celebrated it by going on K.P. This was my last K.P. duty in the army, but there were times when I would have been delighted to have the opportunity to perform this menial service for the chance it would have given me to feed a hungry man.

Our K.P. duties were performed with great rapidity, and then I went with the company to 'police-up' quarters, which included cleaning from the grounds adjacent to the barrack every vestige of rubbish including, of course, cigarette stubs, chews of tobacco, and such,—a task unsavory to all.

Our period of quarantine at Camp Travis was over, and combat companies were now formed out of the recruits from the Depot Brigade. I went to L Company, 359 Infantry. To this company, also, came Bill Goodson, Ike Lawler, and Pete Merk. Joe Templeton went to K company, while Rogers Tipton went to Headquarters company. It so happened I was nearer to Joe Templeton than to Rogers and as a consequence I saw Templeton oftener and Tip less frequently. Ben Whatley was assigned to B Company, but later, when we reached Colmier La Haut, France, he went to a truck company. Willard Newton was sick, as was Jim Alexander. Willard was left at home and later received a commission. Jim came across the water later in the year and of him there will be more to be said here-after.

Work began in earnest when we arrived at L Company. Sergeant Dennis E. Meeks was the First Class Sergeant of the company and W. C. Morrow (called Molly) was the ranking line-sergeant. Two better men could not have been selected out of any regiment. We were fortunate in having with us, also, George Singleton, our old Jacksonville friend. Holcomb, from Rusk or Alto, and Dickerson became signal-men in the company. Dickerson had only one lung. This condition had been brought about from a bad case of pneumonia and a necessary operation. He had contracted pneumonia from standing guard in the rain just after getting up from a sick-bed of measles. Luther W. Thompson was company clerk, and as he had been engaged in similar employment to mine when a civilian, there was a natural bond of friendship.

During this time we received several men from Dallas,—Loeb, Tebbe, and McVey,—real men. Then King came back to us . . . and what an acquisition he was! . . . And Stonewall McKinney, already with this company before we arrived, showed his superiority as a man. All told, we had a fine class of men,—men of intelligence and fearlessness, the class of men of whom any commander might have been proud. Of course, however, we received a few of the chaff, one in particular I recall. Loud mouthed, vile in speech, he was a constant nuisance to all who were forced into his presence. And there was another (strange I should have forgotten his name), who was an ex-convict. He was callous, lazy, ignorant, sulky, and mean, and with an advantage he would have been extremely dangerous to any one who attempted to exercise authority over him. An experience with him in France was cause for me to remember him vividly, but his memory, like the passing of other pains, is only a treasured recollection.

Day after day we did 'squads right', 'squads left', 'by the right flank', 'to the rear march', etc., until at night the commands, sharp and shrill, rang out in our tired minds. I was not physically fit to do the strenuous work set for us but I determined to stay with it until I regained my strength or succumbed in the effort. Forced, however, to remain in quarters one day through illness I was left behind while the other boys went to the firing-range at Camp Bullis, some eighteen miles from San Antonio. I did not rejoin my company there until the boys had been shooting about three days. I was finally able to make the trip and several of us who had been left behind were transported in trucks. The run was not a pleasant one because we had all our equipment with us, and as it was our first time out with it we knew little about handling it.

At Camp Bullis we had our first real experience in army life. We "lined up for chow", "ate without tables", etc., and of course, we rookies thought this "the worst yet". We could not then understand that this life of a soldier was a real spring-picnic compared to what was to come. We did, however, work from early morning to late at night. At day-light we were in the company streets standing reveille, eating breakfast, and policing up the area, and as soon as it was light enough to see the targets we swung into line and marched away to the 'firing-line'.

As I approached the camp that first day I had a peculiar sensation when I heard the first reports from the many rifles fired by hundreds of men, ten shots to the minute. This was new in the realm of sounds to me! After a few volleys, however, an ordinary man accustomed himself to the noise, and by the time I was shifted to fire-duty I could fire unperturbed and with ordinary accuracy. And with this simulation of war-fare there came a glee un-explainable.

My arrival at Camp Bullis was attended by a rare good fortune. I was assigned a tent with 'Bull' Durham[2], a great fellow. I do not know that I ever knew his first name. Perhaps he should not have had one. Any other than his newly acquired cognomen would have been less descriptive. He was a huge man, with the voice of a bull and a kind disposition. Considerate of all, he never failed to take more than his share of the misery if by so doing

2 After the war 'Old Bull' was elected County Judge of Tyler County, Texas, residing at Woodville, where he and I often spent pleasant reminiscent evenings.

he could help the other fellow just a little.[3]

'Bull' and I struck up a cordial friendship. He had a 'learning disposition' and was the best of company. I can see him yet on these first days at Camp Bullis as he goes lumbering, swaggering, rambling along the company streets, his big feet slapping rhythmically against the cobble stones, humming a tune redolent with strength and courage. Although to some he may have been an amusing sight, to me he was a tower of strength.

When once we had established ourselves in camp,—and it did not take much time to accomplish this,—we began to trod the rough, brambly hill firing at long range and disappearing targets previously placed in the mountains. This was done, of course, to train us in long-range and fire-at-will shooting, and in the accurate, quick use of the rifle. I do not recall my average of marksmanship but I think it was about 85%, which most likely was fair. Bill Goodson developed into one of the best rifle shots in the entire company. Holcomb was a close second, and it was entertaining to watch these two friends as they so earnestly tried to defeat each other.

Before going to Camp Bullis all men had been issued regulation army "webb packs". This equipment-carriage was of canvas, having a holder on the back into which you fastened your bed-roll. In the top was a compartment, or pouch, which held the mess-kit. The knife, fork, and spoon could be placed either in the mess-kit-pouch or on the inside of the mess-kit proper. The pack fastened so it could be swung around the soldier's shoulders, two straps forming armhole loops, which fastened also to the bottom, or tail, of the pack. Another strap branches off from the larger one. It was fastened to the front of the pack-belt in which one hundred rounds of rifle ammunition was carried. A full-pack was supposed to weigh eighty-five pounds, but many a soldier, burdened down with his equipment after a hard day's march, has decried the accuracy of government weights.

While at Camp Bullis the men were severely poisoned by what we understood to be spoiled meat. This may not have been the cause. Most likely the origin of the general sickness was lack of sanitation, probably directly attributable to the failure of the soldiers to clean their mess-kits.

3 (Ed). Bull Durham was, and still is, a popular tobacco for self-rolled cigarettes.

This could easily have happened for there was not always a daily inspection of the kits and it would have taken only a few hours for the refuse in the kits to become contaminated. I do not recall having been sick from any cause at Camp Bullis but I do not think I had acquired any particular art at that time in mess-kit cleaning. Many times we were without hot water and we had not then learned the trick of scrubbing our kits with sharp sand, afterwards rinsing them with cold water.

We had a young Jewish boy in the company at Camp Bullis with whom we had lots of fun. He never seemed to get the significance of things. He was perfectly willing to do, but just what it was all about did not register with him. As a marksman he was a complete "flash in the pan", or as we later on would have been called him,—a 'dud'. One of the officers decided he would check up on the boy and determine why his marksmanship record was so low. They went together into the trench from which the shooting was to take place. The boy fired several rounds and the signal came back from the target each time: "No hit." The officer then said: "What the hell is the matter with you anyway?" The youngster dropped his rifle in the dirt and spreading his hands preparatory to speaking, said: "Captain, I maka as much damn noise as any of 'em, eh?"

As all the men of the company had not finished their problems at the same time, we returned to San Antonio in sections. Abiding my time to return to San Antonio I had an opportunity to loiter around Camp Bullis for a short time during which I formed some friendships which have lasted throughout the years. Having nothing more to do on the range I went to the infirmary where I had my arm dressed which was giving me no little trouble from the small-pox vaccination. Here I met Joe Templeton who was also suffering excruciating pains because of the neglect of the doctors. After getting out of the infirmary we decided to go into the city ahead of the other men and were in the act of hiring a car for that purpose when our motives were revealed to an officer. What he said to us was sufficient to deter us forever from attempting to spend our own money for government transportation. We became impressed with the fact we were travelling under orders and it would be well for us to stay in the camp until the army furnished the transportation even though the war should be over before the trucks arrived. Conveyances finally came and we went back feeling more fit than when we departed.

Back again at Camp Travis we were placed at hard drilling. This was the most trying period of my army service, probably because of its constancy and sameness. Our service in France was severe, but the service over there was always punctuated with something novel and, in addition, we had become hardened to the physical grind.

Hour after hour, day after day, week after week, we did close order formations with a few battalion formations interspersed for good measure, and at the close of each day we would lie down petrified with fatigue. Seldom, in the last weeks, did we get 'Sundays off. I recall, however, one Sunday when I was permitted to go to the city. I called on a friend. After dinner I lay down for a few moments rest. The next I knew I was being awakened. My host was telling me that the day was over; that it was time for me to return to camp. My leave of absence was up. I must have been terribly tired.

Late one afternoon our captain, Berry M. Whitaker, formed the company and, with great solemnity, announced: "This company is going to France. We are going right away. Is there a man here who does not want to go?" No one replied. He commanded: "Dismissed!" I have often wondered if that, company did not have a 'pack of silent liars'. We knew our time for departure from America was at hand. Our training, although short, had been intense. A few more days and the finishing touches were put upon the training of the company. We passed in review before the general. We spent evenings at clothing inspections and fretted at the uselessness of these exasperating tests.

Being sick, again, one morning, I was ordered to the infirmary, and imagine my surprise, upon returning to the company-street, when I learned my father was searching for me. Then, instead of reporting to my company for duty, I went to my bunk and spent that and the following day with my father, knowing my infirmary report would be delayed in getting back. When it did arrive I was ordered back to the infirmary, and upon getting back to my company again I found the men in an intense commotion at the barrack. The entire floor was littered with straw and bed-sacks of the men. My 'belongings' were scattered all about. Soldiers were packing furiously. I sensed it: We had been ordered to move.

I had little time in which to roll my pack before the whistle blew, and the sergeant called: "Outside!" Down the steps and into the streets we scurried. The captain appeared before the company and announced: "We are leaving for a foreign country."

There is something weird about the acts of people when they prepare to take the first step into the unknown. An almost invisible pallor seemed to pervade the ranks. Groups of visiting fathers and mothers, standing nearby, talked inaudibly amongst themselves, but all the time their eyes were upon their sons standing in the ranks. Men spoke hesitatingly to each other as they stood listlessly awaiting the next move. And then the spell was broken. King unslung his one-string guitar and thumped:

"You're in the army now;
You're not behind the plow;
Ta-da-de-da
Da-da-de-da,
Ta-de-de-de-de-da."

Captain Whitaker passed along the company eyeing each man as he progressed until he reached the head of the company where he stood before a comely young lady. By her side was a man. There was a resemblance and I concluded that father and daughter had come to see the departure of her hero and his men. There was a shimmer from her hair as the sun appeared for a moment. The captain spoke to her intently. His hands were twitching a nervousness and his face, always red, heightened in intensity. A silence fell. He stepped to the head of the column. "Company, attention! Forward, march!" and,—as he passed her,—his eyes went to the left, and with hand salute, he said: "Goodbye . . . I hope I see you again." We were off to the rail-head.

We soon loaded into Pullman cars where our comfort was far beyond our expectations. We travelled three men to the section, and as this was an immense improvement over the conveniences of Camp Travis we were in rather high glee. All of us had become weary of the life at Camp Travis and welcomed almost any innovation. The irksomeness of the Camp Travis life is illustrated by Bill Goodson's attitude some days before we rode away. We had been enduring the constant drilling, passing-in-review, and other man-killing formalities, until we were ragged and worn. Bill and I were talking one evening when he in all seriousness said: "Chris, I understand we are leaving this place. I am glad we are. It could not possibly get any worse." To this, of course, I agreed, thinking there was not a conceivable condition which could be so exasperating,—but these words . . . at a time when both Bill and I could see the humor of them . . . were thrown back upon me, months later . . . and then, truly, Bill and I took a laugh when to do so meant the risk of our lives.

Before leaving Camp Travis I had abandoned all idea of being transferred. I was informed that Major Barnes had requested my transfer, as he had assured me he would, but the application had not been approved. It was many months later that I saw the notation made by Captain Whitaker on my request. It read: "Disapproved: I want to keep this man in my company for a non-commissioned officer".

L company received four officers some time before we arrived in England. My mind has never been clear as to the time. Lt. Victor Nysewander was the ranking lieutenant. Lt. John R. House from Yoakum, Texas, soon ingraciated himself with the entire company. Lt. James A. Baker Jr., of Houston, Texas, lived to go with us through many trials and to merit the esteem of all survivors of L company. Lieutenant Montgomery Fly led his men into the hell of Saint Mihiel, and died after being grievously wounded, refusing to be evacuated until men worse injured than himself were first cared for.

We had vacated our Camp Travis barrack on the morning of June 5, 1918. An I. & G. N. train took us back over the route I had travelled toward San Antonio on April 26th. We went as far as Taylor when it was discovered one of our men had gone A.W.O.L. Consequently, an armed guard was placed on the train, and strange to say, I was one of the guards. I was placed as sentinel on the "kitchen-car" where I remained, with regular reliefs of course, until we arrived in Jersey City. The duties as sentinel on this particular car, however, carried with it some compensations. It had put me where,—should I be so inclined,—I could separate the United States War Department from goodly portions of the edibles intended for the other soldiers on the train.[4]

Bill Goodson and I were bunking together, and thanks to my taking disposition we were well supplied throughout the trip. Sergeant Morrow . . . dear Old Molly . . . Joe Lloyd, commonly called 'Lloyd George', and George Singleton occupied the section adjacent to me. An inter-seat division of spoils resulted, and these sergeants were scrupulously careful that Bill and I got no extra-duty assignments.

My disposition was not entirely selfish, but of course I could not, for fear of detection, permit a promiscuous distribution of the stolen edibles. However, I let no chance opportunity pass for others to do as I had done.

One of the men came to me after we had passed through Palestine, Texas, and evidently not knowing me very well seemed to have a hesitancy in making his wants known. Finally he inquired if there was any chance to get an orange which he could see in the car-door. I replied to him: "You had better not let me see you get anything out of that car." Whereupon I turned and looked out of the vestibule door into the inky darkness of the night apparently enjoying the passing scenery. When I turned back to the car again the number of oranges in the box had diminished most materially during my nocturnal inspection . . . and I could say in fact I had not seen him take anything from the car.

We arrived in Jacksonville about 12:30 a. m. There was a great concourse of people awaiting at the depot for the arrival of our train. How people became informed about troop movements and what troops were on particular trains was difficult for me to understand. As we dragged slowly through the town, I recognized here and there familiar upturned

4 (Ed.) The I. & G. N. train was a part of the International & Great Northern Railway Company, a Texas line that existed from 1873 to 1956.

faces. A wave of the hand, "Goodbye", and Jacksonville was behind us, and I became cognizant of the fact that all people who might be interested in me had been left behind. I had now, in fact, become a cog in the great wheel, one human cog which was to assist in moving backward the great piece of military machinery which had been thrust forward into France, into the very vitals of civilization. As a living, breathing human being, I had no value. As one of the component parts of a rapidly assembling military machine, I was valued at the cost of replacement,—replacement with another human-cog. Then and there I resolved that I would care for myself, and that I would place upon myself a value beyond replacement. I sincerely believe that it was the precaution springing from this realization of 'soldier-value' which prevented me from being one of the boys now assisting in pushing up the poppies in Flander's Field. It became an axiom later on that he who did not have a personal interest in himself would first fall under the hardships of war.

In giving myself, however, that consideration necessary to preserve me from the pains of exposure and danger, I cannot even now think I neglected to do those things necessary to be done by the faithful soldier. Neither did I so conduct myself as to impose my burdens upon others. I kept in mind some of the essentials, such as securing as much rest as possible, keeping dry and warm—when I could—and above all thing, I tried to keep clean. I learned to value my teeth. I learned to keep them clean under the hardest conditions,—a common neglect amongst soldiers. And although I enjoyed many hours of relaxation through the use of France's famous wines, I tried to keep ever present in my mind the dangers to which one exposes himself through over-indulgence. From the very beginning of our soldier-lives it was evident to me that the chief danger lurking in our paths was: association with women. Solving that problem in its only possible way, I remembered the warning:

"Poor race of men!" said the pitying Spirit,
"Dearly ye pay for your primal fall"—
Some flowerets of Eden ye still inherit,
But the trail of the Serpent is over them all!"

By early morning we were in Texarkana. Here we detrained, took the usual soldier "setting-up" exercises, and paraded in the streets. Just before entraining again, I saw something which will forever remain with me. The father of Sergeant Morrow came to the company to tell his son "Goodbye". Morrow was in charge of the company, and while we rested a few minutes the father and son talked. Suddenly the father kissed the boy and with tears streaming down his face, turned away. Bill was standing near me. I asked him what he was thinking. He said: "The same as you: Will he see that boy again." He never will!

Before leaving Texarkana, one of the men put his rifle to an unintended use. He went into one of the toilets of the depot and shot himself through the head. This trouble might have been saved this despondent youngster had he only known with what willingness any German would have done the same for him. Peculiar as it may seem a suicide in the company created little comment. The men took it as a matter of course and the boy and the incident were soon forgotten.

Our trip was over the route extending through Little Rock, Nashville, Bowling Green, Louisville, Cincinnati, Dayton, Akron, Warren, Greenville (Pa.), Johnstown, and Jersey City, from which place we ferried to Long Island and rode the train to Camp Mills.

We had a disgustingly unpleasant stop-over at Little Rock. The men were getting travel worn. We wanted to be relieved of the tedium of military regulations. A city lay before us and the desire to see it, especially as it became known we were to be there several hours, became intense. The stern military orders, however, forced us into formation. Back and forth over the rough railroad yards we marched. They would break our independent wills. We must learn to lay aside desire and knuckle to military orders! And the drill was the more irksome for we had become stiff from train riding, and we left the city with resentment against all. But at Bowling Green, Kentucky, our fortune forced our spirits higher. We left behind our dudgeon and marched with ever mounting exhilaration toward the main part of the town being cheered onward by a near ovation from the people who lined the streets. At a park we were dismissed and learned their ovation was not to be confined to the 'vocal',—as is so often the case. We received as evidences of appreciation from a grateful people the most palatable of life's delicacies. The inner man was filled to a satiety and soldiers from the Southland wended their ways back to their train surfeited with food and cheered in spirit by the contagious patriotism of a sincere people.

The mental state and physical condition has much to do with an appreciation of the artistic. This I realized after sinking back upon the Pullman cushions leaving Bowling Green. Flitting scenery had a meaning to me, and I was almost ecstatic in my admiration for the "roughs" of Kentucky. A disheveled umbrage concealed the roughhewn conformations of nature, and as our train wended here and there, up a mountain-side, down a glade, I fell to a contemplation of the beauties of nature, to marveling at the accomplishment of man in his efforts to build his roads and highways through these places. Feats almost inconceivable! But I was soon awakened from my reverie by the caustically rasping call: "Fall out at next stop for exercise".

Our train slithered to a creaking stop amid escaping steam and raucous cries to "Fall out . . . and make it snappy" We were at Latonia,—the outskirts of Covington, Kentucky,—the renowned race-tracks of America. I had visions of sleek horses and cheering multitudes . . . but . . .

The day was hot . . . and they marched us onto the race-course where shambling feet scuffed up pungent, noxious effluvium from the dusty track. Our nostrils rebelled at the rancid stench, and breathing became almost an impossibility. Officers stood back . . . out of the lazy, spiraling, fetid, feculent-laden dust . . . and with staccato sharpness hurled their commands at our marching columns.

No one expressed regrets when we moved northward again. The day had waned before we reached the Ohio River. In the distance could be seen the shimmering lights of Cincinnati. . . . The train came to a stop. Bill scanned the impenetrable darkness and stepped downward. His feet slipped off the ends of a tie . . . Down . . . down . . . he slid . . . He grasped desperately at the side of the train . . . and a soldier caught him, pulling him up again . . . Only by inches had he missed falling into the Ohio River over which our coach had come to a stop. A full realization of his narrow escape cast a funereal pall over my spirits. I had almost lost my friend . . . and although the towering buildings of the city loomed menacingly over our train, I permitted the fall of night to obscure it from my view, and when I awakened we were in Greenville, Pennsylvania.

I have heard many noises in my life-time, but the most stupendous noise,—with the exception of the gun noises later to be heard on the 'front',—ever to greet my ears came to us as we entered Greenville. It seemed all the whistles in the city were blowing, and countless bells were clanging, jangling their discordant, dissonant din. The people were rampant with a feverish exaltation, but their cries died still-born in their throats, unable to penetrate the metallic pandemonium. Cheering, yelling people crowded the car-sides and a veritable avalanche of cakes, magazines, cigars, cigarettes . . . everything imaginable . . . was showered upon us.

And when I looked upon this fierce enthusiasm, when I came to know we were not especially in their graces . . . just soldiers passing on our way to duty . . . my mind went back to my Southland. I could see no such fervor there. The fierceness of these people of the East had far outstripped our timorous enthusiasm. We of the conservative South could not comprehend with what seriousness our brothers of the East were taking this war. To these people war meant an accompanying adoration for all soldiers who carried it on. Contagious as was their spirit, I could not be imbued with their flaming, flamboyant intensity sufficiently to prevent retrospection. Had we failed to comprehend the meaning of this war, or was there a climacteric difference in the people of the South, North, and East which ushered us into this maelstrom? Were our people jocose and reticent whilst these consumed themselves in their fervor? Or was there, possibly, another reason. Did not this war more poignantly effect the everyday livelihood of these people of the North and East than it did ours? I bit off the end of some war-enthusiast's fragrant cigar,—a tendered token thrown through the window,—and sat back to meditate, as the din, now growing fainter and fainter with each turn of the wheels, came to me as a far-off inanimate thing . . . and I concluded that, probably, this was industry's war after all!

Our trip across New York State was uneventful. Johnstown, (Pa.) a pleasing place, brought back to my mind the woes of its flood . . . but I, too, had now become engulfed in a flood . . . perhaps a flood of fire.

We rolled past a magnificent cemetery, situated on a hilltop, where lay the dead of a century. Sturdy granite posts at the entrance held, outstretched a banner, the handiwork of a wag:

"Wake up! . . . Your country needs you!"

As we entered Jersey City the order was passed to the guards to stay on post until commanded to join our companies. Someone 'slipped a cog', failing to order my relief, and when the company scrambled off the train, I did not 'scramble'. My company formed in the railroad yards paralleling the coach in the vestibule of which I stood. Roll call was had. My name was called and I answered 'from a distance'. The captain sent a non-com after me, and when I arrived he streaked the air with blue because I had 'held up the company' . . . but I was in the army now, so why explain anything! I got my pack and marched aboard the SUSQUEHANNA.

The Susquehanna.

Our sluggish, slothful boat, teetering almost to the water's edge under the motion of overcrowded soldiers, sludged slowly into mid-stream and pointed its prow toward New York. The massing of human bodies evidenced itself to those of us who had been caught on the inside of the boat. A mephitic, noxious odor drove me to 'windward' where I drank deep of a cooling breeze and picked out the 'Old Suspension Bridge' toward which we were plying. A sparkling new bridge loomed over us, and we pulled from the main stream toward the island on the right where we docked. Our trip had been short.

Again there was a blare of bugles, a clanging of chains and the flump of the falling gangplank. Scurrying feet responded to raspy commands and long files of men thumped their hobnailed shoes on wharfboards. Before us stood a train of 'yellow' cars.

The noise died down as our train wheezily pulled out toward Camp Mills. There was a long walk with heavy packs biting into shoulders and trickles of pain running up and down our backs. Camp Mills showed before us,—tents, long lines of them, with an occasional wood building. Here I, for the first time, saw an army-prison built in the open. It was a formidable thing, something to be thought about. The main prison was a heavy wire enclosure, said to have been electrically charged. But I do not know. I did not test it. To the north was a 'lean-to' shack in which, evidently, the guards were allowed to sleep. Just a peep through the wire convinced me the prisoners were not having the best time of it. The wire-enclosed pen was fairly filled with men, most of whom sat upon the ground nursing a grudge. They appeared to be a hard lot,—disheveled, unkempt, bedaubed. To me, then, they were a hard lot. They may have been so in fact, but later experiences with men and officers in the army convinced me that army prisons incarcerated about as many 'good fellows' as 'hard-customers'.

As we passed these unfortunates some of us, in our thoughtlessness, jeered at them. Immediately, and with a spontaneity springing from their contempt, we got what one of the boys called "a heluva good cussin". We concealed our chagrin by laughing at one of our men who shouted: "O, boy! How I aspire to the delicacy of their art." . . . And we passed deeper into the camp diverting our attentions to many airplanes flying low over the camp, passing well-stocked warehouses, and arriving at a city of tents. We slipped from under blistering packs and sat upon the ground.

I was assigned a tent near the most westerly end of the company street. The day must have been a rather busy one for I became conscious of the fact that night was approaching by the time we were dismissed. A shift of the wind to the north-west accentuated my consciousness, for with the wind there came a mighty stench, a fetor which raised itself to High Heaven and lashed at our nostrils. I knew I had drawn the 'latrine-end' of the street for my abode. A general clean-up of the place, however, relieved the situation, and I became a convert to the efficacy of lime.

We remained at Camp Mills eight days. The camp was a monotonous one. Lying flat on the island it was wind-swept and sand-covered. The shifting winds plastered a grime over our clothing. It found its way into our hair and even our food was gritty. Then, each day we were required to stand 'clothing inspection', and this malevolent condition became a piquant one, for the sand had permeated everything we had, and the officers complained bitterly about the condition of our 'clothing spread', a condition no man could have prevented. Stoical men now recognized the injustice of the criticism but there was nothing we could do about it . . .

"You're in the army now
You're not behind the plow;
You'll never get rich
By diggin' a ditch
You're in the army now."

As our actual training period in America was over we had no duties to perform after 'Retreat', therefore, each man developed a technique in 'running the guard' to go to New York, Coney Island, or other nearby places of amusement. It became a rather difficult thing to keep a bunch of soldiers within a camp-area when New York, only a few miles away, stood beckoning with her sinister mysticism. Coney Island held a lure for those who would drown the realities in gayeties. And I applied for a pass to New York which was refused. But I went.

Running the guard-line at Camp Mills was not so easy of accomplishment as it was at Camp Travis, for the soldiers standing guard at Camp Mills were not members of our division, and with a soldier's duty to perform and having no personal interest in the men of the 90th Division, they were less tractile to our wills. But . . . we had to go to New York . . . even though we had to "seem a saint, when most I play the devil." And imagine my surprise, even suppressed consternation, when sauntering along one of the streets of New York I came face-to-face with two officers of my company. I knew the orders to remain in camp applied to officers as well as men, and taking advantage of this information I saluted and passed on. No mention was ever made of the incident by either officer, and I chalked them up as 'good fellows.'

Seductive tales of nights at Coney Island trickled back to us from other soldiers. This bewitching Seductress of the Sea would have us drink from Neptune's cup, and Templeton and I slyly slipped the company street, slithered past an inacquiescent sentinel and sat that night before a succulent evening repast in one of New York's famed cafes. And when we paid our bill we wondered if there would be enough remaining for Coney Island.

A swift ride in a clattery, clanking coach brought us to the midst of a throng of people, who milled around to a tawdry, tuneless, singsong mechanical effort toward music. Hawkers raised their voices above the din to invite the unsuspecting to their "skin-games". Some tarried at their doors, but we passed on with the crowd to a large dancing pavillion at the water's edge. "Taxi-gals", 'girls of the metropolis', left the dance-floor to mingle with the new arrivals. A hand laid tantalizingly on an olive-drab arm, a suggestive nod of the head, and another 'taxi-gal', all "bound round with a pair of lovin' arms", was dancing across the floor . . . and headed for another night!

I walked up the beach in time to see a luminous moon slip from behind a bank of clouds, and scurrying figures arose along the water-front retreating from soft sands to be enveloped by the crowds. Coney Island to us was a failure, so Joe and I decided it was time to look over the 'Big City' again where we arrived by riding the elevated railroad. We had our pictures made, as all soldiers and most other people do on such occasions. These we sent back to our Texas friends. We now realized the night was almost gone and began trying to get back to a depot where we might board a train for Camp Mills. We met a man on the street and from him asked directions. He told us to go into the next block where we would see an entrance to a subway tunnel. 'Green' as we were, we could not find the entrance, but we continued to search for it until we came to the same man, whom we did not recognize. He, however, recalled us when we asked him the same question again. "Shall I have to rub your noses on it before you can see it?" he said laughingly,—and well it was that he had laughed, for by that time all soldiers were immune to the consequences of physical encounters. We had been taught we were fighters. We were feeling we were such in fact, and the slightest provocation would have resulted in trouble. The kindly disposed man, however, chided us about not being 'city broke' and led us to the underground passage. Wishing us a 'safe return', he left us and we continued our journey into the "rat hole", as Joe called it. We found ourselves in a large room. A stile pointed the route to the 'underground railroad'. Placing a nickel in a slot we went through the stile and into a car which was filled with people to the very doors. We knew as little about getting around under-ground, of course, as we did about finding the 'rat-hole', but we had learned to ask questions, so as we passed the conductor we told him we wanted to get off to catch the car to Camp Mills. He said nothing, but after we had ridden for what appeared to me to be an interminable time, the conductor called out: "Hey there! You two soldiers, and you sailor! Your stop for Camp Mills." I could scarcely believe it, for he had been putting off and taking on hundreds of people since we had spoken to him. My admiration for a New York streetcar conductor was enhanced most materially when we alighted to find we had arrived at the proper station for our departure to camp.

We had alighted in the Grand Central Station. I had previously been there. Joe had not. The assembly-room into which we had come was underground. Joe made the natural mistake thinking he was on the surface. He suggested that we go out on the street, and banteringly, I told him 'to lead the way'. He tried one door after an- other until, finally, I laughed at him, and told him he was underground. Then he got 'all riled up' and 'lowed as how I was a liar', but as it was just old Joe talking I did not object to the inaccuracy of his fervent expressions. To 'cool him off a bit' I lead him into an elevator from which we went out onto the street. New York was dark that night with only a few lights showing here and there. The people were afraid of air-raids and an order had gone out to cut out the lights. Americans now began to expect anything out of Germany. As there is nothing pretty about a big city at night with all the lights out, we went back into the station and caught the next train to Camp Mills.

We arrived at Camp Mills via the 'back route' and succeeded in slipping past the guard just in time to hear the call of the bugle. We ran fast and got to our tents in time to get under the covers with our clothes on before a sergeant came around to check up on the company. 'Reveille' blew and I fell out to formation without a wink of sleep the night just passed. But we had had a good time!

The day after our trip to Coney Island was a strenuous one. It would have been worse had it not been for the sympathetic understanding of our Lieutenant. We began the day with the usual clothing inspection, after which Lt. House took the company for a hike. The news traveled through the company that we were going to Oyster Bay, and that we were to see former President Roosevelt, who was attracting much attention then by daring President Wilson to permit him to take a Division to France, saying that he would go and guarantee not to return. This threat at the slow-moving War Department, 'ham-strung' by politics, made Roosevelt a double hero in the eyes of all soldiers. I wanted to see him but I was terribly tired and my enthusiasm was cooled materially by my physical condition. Lieutenant House would let us walk a short distance and then permit us to lie down. Each time I sat upon the ground I went to sleep. I think he saw this and the intervals of rest continued to get longer and longer as the day went on. We did not go as far as Oyster Bay, and, of course we did not see the former president, but we were dismissed earlier that day than usual. Consequently, long before darkness had come I was doing "bunk fatigue" and enjoying every second of it.

All in all, our stay at Camp Mills had not been a disagreeable one. I had seen a little of the Big City, had amused myself watching the eccentricities of the milling crowds at Coney Island, and had recuperated to some extent from the arduous drill back at Camp Travis. But the eighth day of our stay came, and just as the sun was setting in a murky sky, orders came 'down the line' to "fall out at the west end of the company street", the 'latrine-end' of the street as the boys dubbed it, and some wise chap prophesied, "Now. there's something dirty up." The captain came . . . and what a formation! There was never another one like it in any army. He called upon us to form a circle around him, and almost in whispers he told us we were to sail tomorrow,—"a foreign land", he purled with bated breath!

"First Call' came at 2 a. m. the following morning. The men knew it was the 'Last Call' in America. They went about their duties with a brisk stoicism. There was no grouching, no complaining. Once in a while a little jocular banter, perhaps a little profanity, passed from man to man, friend to friend, but the departure preparations were business-like and thorough. If there were any intentions of desertion, none became evident. That seemed a little strange to me then, for it would appear if there were to be desertions they would come at the last moment. Human nature is such that the weak break just before the test. The condemned man commits suicide just before execution of sentence. The coward turns to flee before his assailant strikes. Desertions are always expected in the army. Very few actually occur, but the greatest test comes under the strain of uncertainty.

When daylight came we moved down the company-street and stacked arms near the kitchen. Here occurred a travesty upon the efforts of the American people to skimp and save for the soldiers. On the day prior to our preparations for departure from Camp Mills the quarter-master-truck had unloaded a large supply of food at our company kitchen. The supply proved to be far in excess of the requirements of the company.

Probably there had been a mistake in the order. That I cannot say, but the fact remains. There had been a rumor running persistently through the company that our captain had fallen victim to the wrath of the major because he had permitted L Company to leave Camp Travis without cleaning up the barrack. In truth, the L Company barrack was in a terrible muss when we left it. I had been one of the last men to leave the barrack. It was then littered with hay, bed-sacks, and rubbish of every kind. Rumor had it that the Camp Travis inspecting officer had wired a vitriolic report of this condition to head-quarters, and now, apparently, there was not to be a repetition of the criticism gone before. Orders came to our company that Camp Mills must be left clean, and the order was complied with, specifically, by our commander. All the food from the kitchen, all the surplus supplies previously unloaded at the cook's tent, every vestige of rubbish,—everything,—was heaped upon a fire and destroyed before our eyes,—and this, even though the men of L company had eaten only one sandwich for breakfast. Our stomachs felt the pangs of hunger, but we saw forty-eight loaves of bread, a quantity of good meat, several boxes of Post Toasties, and many other things, piled high upon the pyre. We knew the American people were stinting themselves to the utmost so that we, Americans, French, and British alike, might have the necessities with which to carry on the war. The world knew the Armenians were on the verge of starvation, but . . . there must be no criticism . . . the area must be clean. What an army! What a travesty! What an outrage against common sense!

We piled into trucks. The fire had reduced to glowing ashes. My spirits had dropped with the dying embers . . . "Save . . . save . . . yes, save for the boys over there." No one in authority tried to explain. No one was obligated to give us an explanation. We were just soldiers, but men would talk, and when soldiers talk they are caustic in their criticism. "Why not pick it up with a truck" . . . "Feed it to us" . . . But what we saw there on that early morning was nothing compared to the waste to pass before our eyes in the months to come.

Our trucks again stopped before a train which whisked us toward New York and the docks. We unloaded to sit down and rest, but there was a Red

Cross canteen at the farthermost end of the building where the soldiers were being fed, and I slipped back into line and took a proffered sandwich and a cup of chocolate. There was a serious mien about these Red Cross women as they went about their work, methodically though dexterously. Someone was handing out cigarettes. The men were chanting:

"Ashes to ashes,
Dust to dust:
If the camels don' get you,
The 'fat-mammas'[5] must."

I took a cigar and would have moved to a seat, but an officer poked a card into my hand. "Sign this, soldier. Lively now." I read the simple words: "I have landed safely in the A. E. F." Well, they seemed to know more about myself than I did, but if they were being prophetic, I hoped they knew what they prophesied. It did not quite make sense to me at first, but when I saw they were collecting the cards I knew they intended to mail them only if the boat landed safely overseas, thus leaving the impression with the homefolks that the card had come direct from across the waters.

5 The Soldier's term for 'Fatimas.) (Ed.) "Fatimas" were a popular brand of Turkish-style cigarettes, as were "Camel" brand.

EDITOR'S NOTES.

Christopher Emmett almost missed his calling as a Soldier as he was a few weeks from being too old to be drafted in 1918. After discovering that he drew the draft number 13, he settled his affairs and prepared to become a Soldier. On April 26, 1918, he joined a group of newly-drafted raw civilians that boarded trains in stations across Deep East Texas. The young men were cheered at each station and at every stop more eager men said their goodbyes to family and friends before jumping aboard the passenger cars. Upon arrival at Camp Travis outside San Antonio, they reported for duty as the Army's newest recruits. As the division history stated:

> The recruits had followed very literally the instructions not to wear to camp any unnecessary clothing. The weather was warm, and some complete uniforms consisted or nothing more than a pair of heavy boots on sockless feet and the all-embracing blue overalls. The amount of baggage varied from a toothbrush to a steamer trunk.[6]

Typical group of "rookies," or recruits, during their stateside inprocessing during World War I. While they are still wearing their civilian clothes, each has his induction and processing card pinned to his jacket or shirt. Stieghan Collection

6 (Ed.) Major George Wythe, *A History of the 90th* Division (The Ninetieth Division Association: San Antonio, 1920), 6.

They were herded like cattle through the large building known as the "bull pen" for physical examination and inoculations. The latter made most of them sick and weak for days. After a few written questions, the Army decided what branch of the service they were best suited for. Within days, those selected with needed skills answered to their names in formation and marched off from their new comrades and away to their new specialties and units to other parts of Camp Travis.

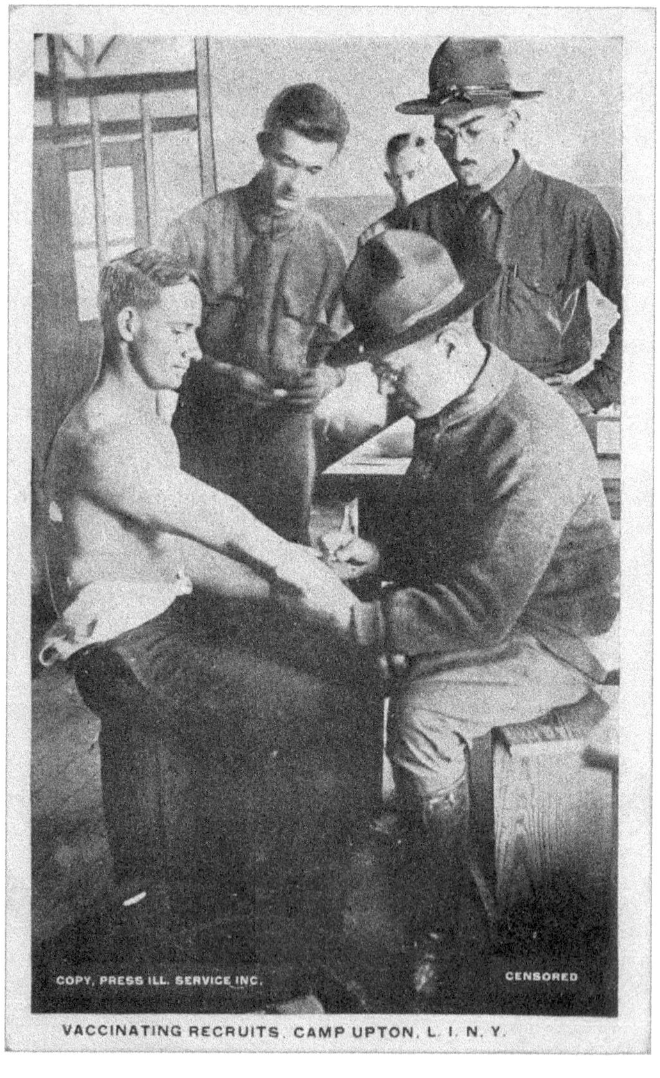

COPY, PRESS ILL. SERVICE INC.
CENSORED
VACCINATING RECRUITS. CAMP UPTON, L. I. N. Y.

An original postcard showing a recruit receiving an inoculation from a medical staffer at an induction center at Camp Upton, Long Island, New York. Stieghan Collection

They learned to don a uniform and adjust their equipment to wear on marches out to the rifle firing ranges. After learning how to load, fire accurately, and clean their rifles, they returned to their barracks to prepare to head directly to France. Just six weeks after arrival as civilians in San Antonio, Emmett and his comrades again boarded trains as Soldiers on June 5, 1918, for the journey across America to board transports for France. They were rushed overseas to be trained by experienced Allied instructors in the new ways to fight industrial war. They arrived in Great Britain and France ready, as their commander General Pershing demanded, to be able to shoot and salute.

A group of new Soldiers in a camp in the United States receiving training outside from information painted on a board mounted on an unpainted temporary building. There were few classrooms available that would hold large groups of rookies, so their training.

While the 36th Division was considered a Texas and Oklahoma outfit, the 90th Division more accurately holds that distinction. The 36th Division began as federalized national guard units from these two states shortly after the United States entered the war. Therefore, they mustered in as understrength outfits in the earlier style 100 Soldier Infantry companies and continually lost leaders as they were transferred to form cadre to help create new units. While originally composed of national guard troops from the two states, the companies were filled to over 250 men each with draftees before sailing overseas. While much of the cadre and leaders were from these two states, the majority of the troops who served overseas with the 36th Division represented almost all the states of the Union.

The 90th Division was a "National Army" Division composed of draftees. These men waited to be called to serve when the Army had enough uniforms, rifles, training posts, barracks, rifle ranges, and sergeants to turn them into Soldiers and send them overseas for advanced combat training. The Texas-Oklahoma Division, also known as the T-O, or "Tough Ombres" *(sic.)* Division, was initially composed of men called exclusively from the two states. In fact, the two Infantry Brigades of the 90th Division were almost entirely recruited from one state each. The 179th Infantry Brigade was raised from draftees from Oklahoma and the 180th Infantry Brigade was composed of men called from Texas. While the 90th Division first saw combat at St. Mihiel on September 12-15, the 36th was first put in the line during the Meuse-Argonne Offensive two weeks later. Both divisions fought well- and suffered heavy casualties.[7]

Reliance on the rifle would restore the power of maneuver and end the deadlock of the trenches, General Pershing believed. His instructions to United States officers training troops were emphatic on this point: "All instruction must contemplate the assumption of a vigorous offensive." He never wavered in his conviction that trench warfare must someday give way to open warfare. Pershing kept pointing out that "in the American army the rifle has always been the essential weapon." At his insistence, the American troops were taken out on the rifle ranges as frequently as possible.[8]

---

7 (Ed.) Wythe, 6-7. Captain Ben H. Chastaine, *Story of the 36th: The Experiences of the 36th Division in the World War.* Harlow Publishing Company, (Oklahoma City, OK), 1920;

8 (Ed.) Richard O'Connor, *Black Jack Pershing.* Doubleday and Company, Inc., Garden City, New York, 1961, 189, 210.

Original postcard of recruits at the rifle range in a temporary camp in the United States. Note the coaches on the bottom right and the score-keepers on the upper right. Stieghan Collection

In additional to rifle marksmanship, Pershing required that the Soldiers of the A. E. F. train for the inevitable day that Soldiers of his command broke through the trench deadlock of the Western Front to win the war in the open country beyond:

> It was my opinion that the victory could not be won by the costly process of attrition, but it must be won by driving the enemy out into the open and engaging him in a war of movement. Instruction in this kind of warfare was based upon individual and group initiative, resourcefulness and tactical judgment, which were also of great advantage in trench warfare. Therefore, we took decided issue with the Allies and, without neglecting thorough preparation for trench fighting, undertook to train mainly for open combat, with the object from the start of vigorously forcing the offensive.[9]

9 (Ed.) John J. Pershing, *My Experiences in the World War*. 2 vols. (New York: Frederick A. Stokes, 1931), 1:152.

U. S. Army Officer Candidates training in the use of the obsolescent US Model 1909 Benet-Mercie Automatic Rifle at the machine gun range at Leon Springs, Texas. These young men would be commissioned shortly and many assigned to lead units in the newly-formed 90th Division a few miles away at Camp Travis, Texas. Stieghan Collection.

Chapter 5

UNDER THE BRITISH JACK

The EL PENOR lay broadside against the wharf. It rose so high in the water that our view across East River toward Manhattan and Jersey City was completely cut off. I worked my way through the congested building intending to get a view of the metropolis, but a sergeant's whistle blew and our lines formed again. The gangplank came down from the EL PENOR. A steady flow of soldiers started upward to disappear on deck. At the head of the gangplank sat a soldier at a table checking each soldier by name and army serial number as he passed. We were told to call out our surnames, then our christian names, and to follow these with our A. S. N. (Army Serial Number.)

The line going up the gangplank was moving slowly. The men seemed to have lost their voices. The checker was having difficulties in understanding the names and numbers, even though the rolls had been arranged in the order of the march of the men. Finally I stood abreast the checker and yelled at the very top pitch of my voice: "EMMETT . . . CHRISTOPHER . . . A. S. N. . . . TWO MILLION EIGHT HUNDRED SEVEN TWO SEVENTY FOUR." And Bill Goodson, following immediately upon my heels and catching the humor of the thing, pitched his piercing stentorian voice above the hubbub and far above my effort and notified the surrounding world that he, too, was going aboard! The checker tilted back his chair and looking up at us shouted: "THANK GOD! AMERICA IS SAVED. TWO LIVE SOLDIERS GOING TO FRANCE."

Two red checks were handed us as we passed onto the deck. One represented the number of the bunk to be occupied and the other the mess on the boat to which we were assigned. The check-tickets immediately disclosed the fact that there would be too many men aboard to be fed at one mess, and it turned out that eating was a sort of 'continual affair'. So many men were on the boat that when the last man finished eating lunch it was about time for the first man to eat dinner.

My first impression of the boat was a pleasing one. The upper deck had cabins which looked clean and fine, and I began to feel we had progressed beyond Ike Lawler's "hell-hubbing gang" . . . but I was to be disillusioned.

On and on I went, following the stream of soldiers as they filed down into the ship. Then I climbed to the bottom of the boat on a rope ladder. With pack on back, the ladder swaying to and fro, I landed at the bottom into semi-darkness. Toward the stern of the boat I found my bunk, and fortunately, right over the keel. I could not then appreciate my good fortune, but sank down to make my home there for the next thirteen days, only to learn the value of the location later.

The olive-drab stream flowed all day long into the hold of the ship. Those of us who had been fortunate enough to be amongst the first aboard now found time to look about. The familiar odor arising from the hold convinced me immediately that the British had only recently used her for a cattle-boat but they had not given her a meticulous cleansing before she was put into the transport service. Designed as a cattle boat, and not for the 'storage' of men, some of the essentials of comfort and all the conveniences were lacking. But France needed men, so what did the British care!

The lower deck arrangement for soldier-bunks was very poor. These 'apartments' were of wood and just long enough for one prostrate man, provided he was short. The width was scarcely sufficient for a slim man, and no man dared to lie flat on his back for fear of being wedged in. To add to the misery each soldier was forced to take into his bunk all of his equipment and our 'over-seas' roll had assumed sizeable bulk. But the men fitted themselves into their niches with little complaint, and as a general rule maintained a commendably good spirit. All were looking forward to the experiences to come, and discomforts were endured or forgotten . . . that is, until sea-sickness pervaded the ship . . . and then . . . but that is another story!

I have never known the number of men loaded onto the EL PENOR. 'Dame Rumor' told us we had twenty-eight hundred men aboard. Besides a few casual companies from Georgia, it held the Third Battalion of the 90th Division, which was comprised of companies, M, I, L, and K, 359th Infantry, ordinarily dubbed the Milk Battalion. The hold of the ship was loaded with cotton on which was stored the baggage of the officers.

We pulled away from the dock at 8:20 a. m., June 20th, and in a short time were looking backwards at the Statue of Liberty behind which was a smoky haze,—receding America.

As we passed down the river, and before we were required to go inside the ship as a precautionary measure against spies reporting the departure of troop-laden ships, I received an impression which staggered me. Exactly nine days before, in the early morning, I had passed over the path of our then departing ship. In the same waters we had met out-bound vessels. Leaning over the rails of the out-bound boats could be seen thousands of khaki-clad men. Our "SUSQUEHANNA" was also filled to overflowing, and our men were cheering, yelling, screaming their enthusiasm to their buddies passing out. But the men on the outbound boats were silent. Not a sound came to us from these men over the short expanse of water. Only an occasional listless hand raised to wave a last "Goodbye". Now, nine days later, we were going out, and other men were crossing from Jersey, following our trail. They were yelling, screaming—just as we had done. Now, none of our men spoke. A peculiar psychology! A realization and an understanding, a contemplation of the future, and nine days had made the change!

I lingered on deck as long as I could. The receding land had a limpid haziness now to me, and I became conscious of the fact that my eyes were playing me tricks. I took a long look at the Statue of Liberty and was thinking of Bartholdi, the Great French sculptor . . . the gift of the people of France . . . commemorating independence . . . when I was brought back from my reverie by a pull at my arm. A tobacco-chewing soldier spitting his quid over the rail, pointed: "Say soldier, if I ever get off this trip I never want to see that tall gal there again except from behind," and I laughed and went below deck, wiping the mist from my eyes.

I came out again after we had passed the ten mile zone. A few vessels were yet within view but they were headed toward New York. Our convoy only was going out. We were now plying a northerly course and some officer volunteered the information that we were to pass near Halifax. I thought there was sarcasm in that informative statement, but it was not so intended. We did pass near Halifax where one of our ships was left in port for rudder repairs. I counted them and we had thirteen ships in the convoy, and I remembered I had drawn No. 13 when I registered so long ago at Jacksonville. I felt a little relief when the thirteenth ship put in to port leaving us to continue the journey with twelve.

A convoy of ships under war-conditions is a stupendous sight. Perhaps the state of the mind has much to do with the impressions received. We were thinking in terms of great power, but the sorry spectacle the nation had made in mustering recruits made all of us doubtful of our national power. We were just a little squeamish. Leaving the United States with our thirteen transports we were guarded by one powerful warship and a number of 'mosquito-boats'. A dirigible balloon,—an 'old sausage',—was towed out by a boat while we were near to land, and then the air was filled with hydro-planes which whipped around, ahead and above us, peering into the water for tell-tale evidence of submarines which might be lying in wait for us. The hydro-planes soon returned to the land, but the 'sausage' lingered much longer. Perhaps it turned back the first night. At least I did not notice it the second day out.

The formation of the convoy was that of an immense inverted 'V,' the warship taking the point, the transports taking the sides of the 'V,' whilst the smaller protecting ships, torpedo-boats, etc., skimmed along the outer edges of the convoy. The balloon lagged along lazily to our right, blobbing up and down as it was buffeted about by the wind. The balloon, then the small boats, left us, but the old warship led the pace out ahead. To watch that stalwart battler was to gain confidence. I became a great admirer of that grim old sea-fighter. I never knew its name, but I would watch it by hours. A great duty had been laid on this inanimate thing. I could not think of it in terms of being manned and managed by men. It was always there, a tower of strength, swift in the water, now playing far out in front, then changing its course to "run out the sides" of the convoy. A big strong, playful, inanimate water dog!

Immediately after being permitted to come again onto deck following our departure from New York, Bill Goodson and I were called to the officers' quarters. Here officers advised us we were to serve as guards on the ship and were to remain as such throughout the entire voyage. I was assigned post at the stern on the star-board side, and was particularly charged with the responsibility of discovering any submarines which might be hanging around with evil intent. I had two shifts of duty, the first beginning at 4:00 a. m. and lasting until 8:00, after which I went off duty until 4:00 p. m., standing guard again until 8:00 p. m. This was not unpleasant service. It would have been rather enjoyable had it not been for the misconception of the officers aboard that all men, whether previously on duty or not, must be constantly doing something. 'Duty-dodging', therefore, became an acquired art, and I succeeded in doing nothing (except my regular guard-hitches) from the time I set foot aboard until I tumbled off the boat thirteen days later.

After coming off duty each morning at 8:00, I would take a little 'bunk-fatigue', then seek out old Joe Templeton, who would already have arranged for the 'eats' for the day. Then we would lie on deck (the weather permitting) and 'laze' the day away. Joe found it rather easy to dodge duty and we were, therefore, together most of our wakeful moments. We, however, were not different from all others. Duty-dodging was unanimous. Acquiring laziness on a ship was easily done, and when once the men got lazy the sergeants suffered terribly. First Sergeant Dennis E. Meeks soon was having a great deal of trouble getting the men to do any chore on the ship, and failing to get George Singleton to comply with an order he shouted at him: "Damn you, George, you are sea-sick, home-sick, and love-sick, and I hope it's fatal."

Our transport passed into a storm area which lasted two days when we were about halfway across. The waves were so high the guards could no longer stand their posts and all men were forced below. This, of course, befouled the air and sea-sickness pervaded the ship like a scourge. Although our ship did not leak, the wind blew a blinding, salty spray over everything, and a slough of water went into the hold. The bottom of the boat became a stinking, clammy, slimy, sloppy mess. And before the wind ceased to blow there was about eighteen inches of water down below. No effort was made to pump it out and the men sloshed around to their great discomfort, whilst the water poured from one side to the other with the listing of the vessel. Templeton had been sleeping on the extreme right of the boat. Suddenly there came a listing toward the right and with it went a rolling rush of water. Joe was completely submerged in his bunk, and he came up blubbering. Thereafter he occupied my bunk over the keel while I was on duty.

The storm finally subsided, and a great, glistening expanse lay before us. It is peculiar how serene is the calm after the storm. An occasional feeble wave caught a shaft of light and threw back to us a beckoning salute. Drab, sea-sick soldiers came on deck to lie down where a welcome sun could dry their bodies and cheer their souls. But the officers would not have it so. From deck to deck they trod: "Move on, soldier! Move on . . . You can't stand there, soldier. Move on, I tell you!" . . . Back to him came the invariable courteous reply, "Yes Sir, Lieutenant," but some bewhiskered lad would sing softly in a tone sizzling with sarcasm as the officer moved away to disturb the meagre comfort of others:

"There's a silver lining,
Through the dark clouds shining
While the doughboys LIE
And dream their lives away."

Pique was suddenly forgotten in excitement when we sighted a sail-vessel showing two masts. Signals were raised to this curious windjammer, but there was no response. Signals were repeated, but again there came no evidence of acknowledgement. And then there was a lurid flash of flame from out the smoke-stack of our warship. Rolls of dense black smoke shot high in the air and trailed back behind the old warrior as it suddenly lurched from its languor, and with all guns trained on the sailor, lunged forward to meet the sail-vessel. It seemed only a moment until the warship had traversed the several miles separating the ships . . . and then the amusing thing happened. A constant stream of flag-signals slid up and down the sailmast. It was signaling her explanation; she was an English fish-boat which had been driven far to sea in the storm; she was intent upon her vocation and had not seen our signals, consequently the delay in making herself known.

Such an incident, of course, was calculated to excite a convoy during those times. It was known to be a frequent trick of the Germans that they would camouflage a submarine behind a fish-boat and then shoot their deadly volleys into the passing ships.

Our mess was in charge of the British, the owners of the ship. The vessel was manned by Englishmen; however, there were several Chinamen in the crew, some of whom did the cooking, some did the rougher shipwork, such as cleaning and swabbing deck. These 'chinks' were not held in high regard by the Americans.

I do not think the food could have been cooked more unsavorily than was this British-Chinese concoction. We received through this mess our first taste of horsemeat, but this was a delicacy compared to some rancid fish they served us. The food condition grew progressively worse, day by day, until some of our men complained bitterly to the officers aboard. Colonel Cavanaugh, a Spanish-American War veteran, was aboard and this condition was brought to his attention who avowed it would have "immediate rectification" . . . but no results! Day after day this continued and then . . . The British opened a canteen on ship. They called it a sales-commissary, and through experience with it I learned this meant a place where the most exorbitant prices were charged by Englishmen to Americans, who were going over to save their country, for food which these Englishmen had either bought in America at reasonable prices or for food for which the United States had paid and which had been stolen from the soldiers' mess.

Most of the United States soldiers before departure had provided themselves with some money. Now that they were starved, a soldier would pay any price to secure a little clean, palatable food, and American money went rapidly into British hands. Each day the prices mounted higher and higher until finally oranges . . . just ordinary Texas Valley oranges, having a regular pre-war price of one cent each . . . were handed over the counter to the tune of thirty cents per orange. For other things we paid in proportion. This I cannot say fostered a kindly feeling between the two peoples.

One night some of us made the discovery that British officers were baking cakes for themselves from the "excess" in the soldiers' galley. This information came to us from the chinks who began the practice of selling us the excess cakes at two dollars per (small) cake. The 'gang' was called together in solemn conclave. The roll was called, and in whispers we answered: King, Goodson, Singleton, Allen (James O.), Emmett, and others. The casus belli was stated. Probably it was King who had interviewed the Chinaman: that, now, I do not know. We were, however, of one accord. The cake-baking must stop. An investigation party was formed. We were to ascertain what could be done and report to the conclave. The report was an encouraging one for those who enjoyed excitement. I recall no quitters. We decided to rob the kitchen, which the committee said would be easy of accomplishment. It was entirely feasible for a man to crawl over the top of the partition into the bakery-store-room in which the cooked food was placed. To do this without detection we had to block the passage way leading along the wall. Volunteers were called for. King's response was instantaneous. In fact, he demanded this privilege as an inherent right, a right he had wished to assert since the time he first ate a British meal.

The passageway was blocked. Solemn-faced sergeants stood turning back wandering soldiers: "You can't come this way, soldier". King's body-guard walked along with him. We picked him up and threw him over the top of the partition and he disappeared, striking the floor inside the galley with a resounding thud. Just as we released King over the partition a ranking British officer approached the bakery unseen by us, entering from an unseen and unguarded door. And there stood King with his arms full of edibles!

British Officer: "Wat the 'ell you doing 'ere, my man?"

King: (who was never disconcerted in any emergency) "Sir: When do we start this fight?"

The officer, sensing that King really meant what his words portended, side-stepped and permitted King to leave the bakery from the now open door. He, of course, did not forget to take with him a full-arm of his acquisitions.

The Britisher passed the alarm to the American officers and the man hunt went up and down the ship: Find the robbers! King's arrival and our having heard the conversation, although brief and to the point, was the signal for a scurry. I almost dived down the ladder into the hold of the ship, where we foregathered on top a table. We had not expected an early arrival of the officers and we were quietly sampling and distributing the loot when shining boots and jangling spurs started down the ladder just above us. To sit there was fatal. We flew, the spoils going with us. I ducked around a corner where I found some blankets hanging from the ceiling in the center of the room. They had the appearance of single drapes, but they were doubled, and between those folds I slipped. I stood perfectly still whilst the searching officers almost trod my toes. When they passed, I ducked again, and in a fraction of a minute I was checked 'in bed', and never thereafter was I suspicioned, although the investigation occupied the minds of the 'brass-hats' for days thereafter. And if I had my guess it would be that Colonel Cavanaugh, versed in the ways of the soldier and probably nurturing no kindly feeling for the British himself, took great pains not to know the names of the robbers. But the form must be carried out!

We submarine-guards would often find obstacles floating on the ocean which would have the appearance of periscopes. I presume our nervousness had much to do with our conclusions, for each submarine signal relayed by us proved to be false. But with the alarm would go a rush for the life boats. As we went farther and farther to sea we became accustomed to suspicious objects floating on the water, and no longer was an old barrel blobbing up and down an imaginary danger. The men became quieter and we strove ahead, zig-zagging our course across the ocean.

Late one afternoon when we had traveled so far toward the north that our hours of darkness were very short, there was a series of long, deep, guttural blasts of our whistle. The flag-ship's pennants rose and fell. The chug . . . chug . . . chug of our engines died slowly down and we drifted to a stand-still on a becalmed sea. An officer came to me and said that a special look-out must be kept tonight; that we had stopped for repairs of a rudder; and the convoy was not going to wait for us. I looked ahead at the big inverted 'V' and we were no longer a part of it. My eyes would not fasten to the water around me. They slipped back toward the east, to the fast gathering gloom of night, to the 'old watch-dog' leading out ahead, smoke flowing back from her stacks, spreading a pall over a tranquil sky. And I attuned my ear to the trill of distant music:

"My country 'tis of thee
Sweet land of liberty,
Of thee I sing . . ."

The consciousness of our disabled condition, of our having to rely wholly upon British marksmanship, sent a shiver of nervousness throughout the ship. Men, previously accustomed to taking to their bunks with disappearing day, sat on deck and talked with their buddies that night; but the throb of the engines came again. Reassurance came back to the men, and when I went again on guard at 4 a. m. we had taken our place with the convoy.

Submarine guard duty on the whole was really pleasant, but the monotony of the thing worked to make the minds of all men strive to outdo the officer in charge, and "soldiering" became the order of the day.

Sergeant Robert O. Flynt was sergeant of the guard. He was one of the finest of men, and our impositions upon him were not born of a desire for vengeance but from pure cussedness, the child of inactivity sired by monotony. The officer of the day had permitted the guards to go to their meals on the hour specified by our meal-tickets. The sergeant was required to stand-guard during our absence, but when he found me taking more than a full hour for my meals, he instituted an investigation disclosing facts which reflected no credit upon my veracity. Not being able to work the 'meal-ticket' any longer, I got a 'latrine-order' posted, after which I departed with great regularity and frequency toward the emporium, but the sudden and unexpected appearance of the sergeant, when I was on bended knees calling for 'Little Joe', broke up that tactic also, and I became a regular attendant to duty.

At night the speed of the convoy was reduced and the ships spread out allowing more sailing space for the prevention of collisions. This was necessary for the reason that no lights of any character were shown after dark. It would have been cold on the norther-most part of our voyage had it not been for the fact that we had an abundance of good warm clothing, including a "pancho", a variety of slicker, which later became known to the boys as the "Mellon Special." But being in the cold air was more pleasant than remaining in the hold of the ship where the stench from the dirty water, the perfumes of departed horses, and the odors of overcrowded human beings, stifled you at first, then brought on a sort of toxic mental invalidism. I, therefore, when off duty,—if not out 'sunning' with Joe—put in my time with the stern-gunner's field glasses. He and I had fraternized while on duty near each other with the consequent result that the glasses were mine at my pleasure to scan the sea. I was astonished how many things were afloat on its surface,—barrels, boxes, crates, ship-masts, just a little of everything. Man does not pass across the stage of life without leaving some evidence of his going.

An effort was made by the Y. M. C. A. secretary aboard to provide diversion for the soldiers at night. A room was set aside for this purpose. It had been made light-proof, and consequently air-tight, to prevent our lights being seen by submarines. Within this dark-box of feebly flickering lights, motion-pictures designed to encourage, and boxing-matches calculated to teach the art of defense, were exhibited each night. I availed myself of the diversion a few times but the place was too much for me. I had been accustomed to tobacco smoke all my life, but this smoke-pot fairly reeked with fumes, becoming so thick at times that the pictures could not be seen on the screen. Consequently, my chief diversion, when not asleep, was lying on top of deck watching the dashing, shimmering waves in the moonlight, or trying to peer through the shoreless gloom. We, trusting sailors upon Life's uncharted sea, could but look above to the source of light, and that failing, strain our eyes into the gloom with an abiding hope. My old university friend, I thought, as I repeated over and over again his SHORE LIGHTS,[1] must have envisaged this journey when he wrote:

"So oftentimes on Life's uncertain main,
When, tempest-lashed and wrapt in rayless night
With warring winds and hostile waves we cope,
And, struggling, sink—and, sinking, strive again—
There burst like beacons on our dazzled sight
The lights that mark the shining shores of Hope!"

When near the end of the voyage our mess got down to real 'hard-pan'. The only thing passable as food was tapioca-pudding. With this meagre food supply it was only natural that there should be a reoccurance of mal de mer. Everywhere you looked you would see a soldier draped over a railing, gagging, retching, swearing. One day the sun came out bright and warm, and the cooks decided to air their kitchen utensils. With this in mind a bread pan was spread in the sun on the second deck. The pan was little more than placed when a sick men on the deck above heaved at the sight of the grease-covered food container and all his salivary spewings landed fairly in the pan. The Britisher looked up in amazement, and then

1 Hilton R. Greer, The Spider and Other Poems.

swearing violently, picked up the pan and started back into the kitchen, shouting, "Now damn you, you'll eat out of it just that way. I've washed that pan my last time."

Our northward journey shortened our periods of darkness most materially. One night a soldier sat with me on deck until some time between nine and ten, at which time the day had not even then waned, so we decided to go below regardless of the conduct of the planets. When I arose the following morning about three o'clock to be ready to go on guard at four, I saw we were passing an island, and it was, again, as light as day. I called this same boy to come on deck to see the island, and noticing it was daylight he remarked: "Hasn't that sun gone down yet?"

On the morning of June 28, land again was sighted. We were told it was one of the Shetland Islands. To us then came a reality; we were in a foreign land. The spirit of the men had been admirable. Few had lost their nerve, but with land coming up out of the horizon to greet us, and with a realization that we were 'Over There', the thoughts of all soldiers went back to America, to home, to friends. We were nurturing recollections of the freedom left behind. Perhaps it was a natural trend of thought . . . this making comparisons. It may have been a maudlin sentiment springing from the echoes of the oft-repeated exhortations of patriotism heard before our departure, but, anyway the words of Lord Byron kept scudding through my mind:

"Here's a sigh to those who love me,
And a smile to those who hate;
And, whatever sky's above me,
Here's a heart for any fate."

But . . . perhaps I was just getting scared and was trying to kid myself into a simulation of bravery!

The night before we sighted the Shetland Islands,—or perhaps it was the second night out (Recollections are treacherous after a mind has been burdened with more important things.) We were notified that our

American warship was returning to the United States under cover of darkness, leaving us without armed protection other than our transport guns. A common-sense interpretation of this made us know we had passed into that area of the sea kept fairly clear of submarines. We guards were even more alert that night, for we had learned to rely more on our own eyes and less on those of the British. Submarines, according to the reports which had come to our convoy, via wireless, had been wiping the commerce from the seas, and none of us had a desire to be dipped into that cool, placid water. Submarines were on the minds of all! Take for example, the statement of my genial friend, Bill, who viewed with me a sight to be long remembered. A soldier who had died on the boat was wrapped in the customary blanket and shoved over-board. Bill watched his rather unceremonious departure and then wryly remarked: "Chris, I will never forgive the man who throws me in that cold water." And no man felt predisposed to take a morning plunge because of a scuttled ship!

Nothing of particular interest happened after the departure of the man-of-war until just before the appearance of day. It seems to me I was never asleep. Perhaps I needed little sleep for the 'snoozes' on deck supplied the required rest, and I was on deck before the break of day. Out of the northeast came a myriad of winking, blinking, points of light. True to the army style no preparation in the minds of the guards had been furnished. No information of any character had been given the soldiers. The result was great excitement on ship. Guards reported their observations to the crow's nest. The report came back: "Mosquito Fleet". This meant little to me. Perhaps it meant less to others, but it was interpreted by my lethargic friend, the British gunner, as meaning that with the departure of the United States gun-boat we were to be met near the end of the voyage by a British fleet of small boats, submarine chasers, which they called the 'Mosquito Fleet' because of their diminutiveness. This was solace itself!

Any attempt to convey to one, who has not seen the approach of such lights, a proper conception of their entrancing, mystic beauty would be futile. Had they even heralded danger they would have been enticing. Blinking, twinkling,—signaling of course,—as they arose over the rim of the sea, infinitesimal at first, they became large as they drew nearer to cut our path of progress and go on with us. Had the ocean been suddenly invaded by a swarm of fire-flies, lighting their flight with their phosphorescent glow, we would have had a parallel.

Daylight brought us an even more interesting sight. We had been met by perhaps as many as twenty small boats of the torpedo-destroyer type. Geared for great speed, they raced along the sides of the convoy. Like playful, scampering puppies scenting their prey, they would change their courses suddenly and run far to sea, scouring the surface of the water as they looked here and there, watching for evidences of danger in an infested sea.

One boat drew near to the side of the EL PENOR and I looked down upon its water-washed deck. A sturdy, strongly built, narrow boat pulsating with power and speed . . . and loaded with depth bombs,—dreadful, death-dealing darts of damnation. On her deck were several small rifles of probable three inch calibre, one being mounted at the bow, one at the stern, and the third in the middle of the boat flexible on its turret. The 'mosquito' was very low in the water and the weather was unusually rough. With each dashing wave a spray showered over the entire ship drenching the oil-skin clad sailors who smiled up to us and cheered us with a greeting wave. These boats remained with us until we entered the zone of safety.

Our trip was not marred by an actual encounter with a submarine. We did, however, receive a message from a foundering steamer some fifty miles behind advising us to be on the look-out for a submarine headed our way. We had, of course, our protection against these sea-snipers, but when we neared the coast-waters additional troubles developed,—mines.

Concurrently with mines another war-operation, its counter-part came before our eyes,—that of 'mine-sweeps'. We saw little of this, but sufficient to understand the method. The operation was performed by larger ships than those used in the 'mosquito fleet', and the process was a simple one, not dissimilar to deep-sea fishing on a large scale. Two ships would handle a seine, or net, between themselves. As they plied forward a wide swath would be dragged in the water. The seine would collect, as it proceeded, such mines and floating torpedoes as had been planted in the paths of passing vessels. The bows, of course, of these mine-sweepers were exposed to the greatest danger. To minimize this danger a buffer was suspended before the bow of the sweeper which would lessen the force of the contact with the mine. In this manner the ships swept the coast clear of danger and our transports entered the infested territory with a modicum of safety.

After the Shetland Islands came the coast of Scot- land,—a beautiful sight. Water for twelve days, then land. And land gave us a feeling of security. At our point of approach the land jutted out to sea terminating with a lofty cliff. Perched on its precipitous tip stood a lighthouse. On this outjutting fragment of mountain,—sombre in its early morning shadows,—stood a house,—held high against a fleckless blue sky,—diminutive, clean, white, glimmering in the rising sun. There was a sight to hold the eye! A terrace ran down the mountain side to the very water's edge, the walk-way being marked off neatly by whitewashed stones. An industrious Scot in his meticulous care had taken pleasure in this accomplishment. Sweeping the coast for miles, the full range of my vision, there was not another evidence of life. A small sailboat with its canvas languid in an almost imperceptible wind basked lazily at the foot of the cliff. We had crossed the ocean!

George Singleton, now recuperating from his mal-demer, and again in the good graces of Sergeant Meeks, but yet held enmeshed in the tangled skein of love, peered over the rail by my side, and feasting his eyes shoreward, avowed that should fate be kind to him, some day, he would return to this very spot where he would forever content himself as a light-house-keeper . . . "but", he added, "I think I'll bring her with me."

With the possible exception that the cliffs along the coast of Ireland were not so high there was a great similarity in Scotland and what little we saw of Ireland. Being Irish it was my sincere regret that our ship did not again break a rudder, for I wanted to touch land in the country of my father's nativity. We had no such good luck.

My dislike for the British during the voyage grew into an obsession. Since so many days have now passed, I can assign no good reason for my feelings. Perhaps it was the restraint on the ship, perhaps it was the noisome recollection of the mess that caused me to show contempt for the British an all occasions. I had my opportunity for expression before I left the ship. We had passed down the Isle of Wales, safely into the 'free-zone', when two husky 'tars' came to the stern and flaunted a British Jack into the wind above my head. I believe it would have been impossible for me to have resisted the temptation to taunt them, so I called up: "Hey, Limey, why the hell didn't you show that pretty picture where a Dutchman could see it?" He answered me: "And the bloody Yank would 'ave us killed!" . . . as if it made any difference!

The last day down the coast was my birthday. Our procedure was slow for the minesweepers were delaying our entry into the harbor. I wanted to celebrate the day on land. Assuming single file our ships came slowly in, a 'big broom' leading us, and on the evening of July 1, the outer edge of the harbor of Liverpool was reached.

This city appeared to be situated about eight miles up the river. Land was visible on each side of the ship as we entered the river. There was a great display of gaiety along the water's edge. People crowded the pavilions and danced riotously. Boats, fast small ones, ran down the river to meet us. They circled and sped on again, waving and cheering. Many women . . . it seemed to me almost all were women . . . seemed to rejoice at our coming. Then there was wafted to us the thrilling notes of "America" and the home-sick soldiers went wild, but pandemonium surged into a riot on the land when our band-master shifted to "Tipperary."

One gorgeously bedecked gasoline launch, displaying the Union Jack and the Stars and Stripes, festooned with the glint of varicolored women's clothing, dashed to the very mouth of the channel and turned to follow us in only when the electric buoys winked their 'mine-danger' signals. Their joy was unbounded,—but I noticed no British uniforms in the launch. 'Stove-pipe' hats, black dress-suits, tonsorial sprightliness, but no uniforms! And I wondered: Are these the British "flag-flappers?" Yes, we had them in America, too!

Thirteen days we had sailed, and there was comfort in the demonstration. There is no kindness on a convoy ship, but here was a welcome, and my spirit of resentment against the English cooled.

Just as the sun cast its last rays around us we drifted into a berth at the wharf-edge, being aided by a chugging, snorting tug which labored at our prow. There was a jangle of steel and the anchor hard by my side, as I looked over at the struggling tug, was loosed. It fell into the water with a splash. We were anchored in Liverpool, England.

The men were not permitted to leave the ship on the evening of their arrival, so we loitered on deck until far into the night. Around the dock a 'ratty' looking people shambled about. The 'elite' had waved their flags and gone to their homes. The moon came up to mark the outlines of the houses in the city, and Joe Templeton and I sprawled on the deck and discussed the morrow,—perhaps the past,— but in tomorrow we had a keen interest. The noises around the shipside soon died away. The moon climbed high and hid its face behind a cloud as if to warn us we too should sleep. We had nothing to talk about . . . in the darkness . . . There are few things men can talk about in the darkness . . . and we went to bed. Our sleep was short, or perhaps we had rested better now that the engines no longer pounded in the bowels of the ship.

We were aroused by the clatter of feet on steel and stone. Soldiers were marching. I became conscious of the fact that we had not eaten, but I followed on deck where my ears were saluted by curses. No one was cursing me, but there was a perfect medley of curses. On the gangplank stood a British officer. He was swearing fiercely. His face became redder with each torrent of vituperation. Not until then did I know any Englishman could show such fervor. He was cursing the wrong man, so I thought, for before him stood Colonel Cavanaugh. The colonel was demanding sufficient time to feed his men. The Britisher was saying: "Damn the men! When I issue an order for men to march, then they march! No man can delay me." And we marched . . . down the gangplank . . . into the railroad yards . . . into a street paved with bricks. We had 'sea-legs', and I walked like I was drunk. I was not, I was sorry.

We formed ranks again, and our packs were heavy. "Squads right . . . march!" and down the street we went. The hard red bricks, slick with morning fog, clacked discordantly under foot. Now there was no crowd to look at us. The 'elite' were yet asleep. Just a few people peered disinterestedly from dingy stone buildings. We waved at them and inertly they withdrew their heads. We did not expect this inertia. Nearly all of them were women, too. In America the women would have been joyously enthusiastic!

We made a long trudge across the city. Our hobnails pounded against the uneven bricks . . all the streets seemed to be of bricks . . . and our feet swelled and pained us. The clatter of falling shoes shod with iron raised a din which reverberated through the empty streets.' Ahead was a long serpent of khaki. I looked back but could not see the end. The poisonous snake, I thought, was crawling toward Germany, where it would sink its fangs deep into the intruder upon civilization.

I struggled to keep apace. I wobbled under the weight of the pack. An officer swore at me for breaking step. I struggled again and slipped on the bricks. I fell and my ankle hurt terribly, but I kept moving. As I shambled along the pains darted through my ankle and I became a walking invective. Then I noticed all the other men were also cursing. But ahead was a railroad station, and we filed slowly into the depot. This was better, now, for we did not have to hurry and keep step. The depot was a seething mass, mostly women, many girls, a few boys, and here and there a British uniform,—usually with an empty sleeve. Now we were standing still, and the command, 'rest' came down the column. I broke ranks and others followed. Lieutenant House smiled. He knew how to smile, and the men liked him.

In the center of the building was a news-stand. I yearned to read. I offered the news-girl money,—real money, American silver,—and she laughed at me . . . But she would 'exchange' some money for me, and she did. I tried to count my money and realized I did not know how. Some obliging lad volunteered to count it for me. He said she had short-changed me. I suppose she had, but what was the difference? I had been 'short changed' too often in the past months to care. I now had money that I could spend, and I was satisfied. I bought a morning paper,—a habit of a life-time. I turned to the court-reports and I laughed to myself. Reverting again to a habit of years! What interest could I possibly have in the courts of England? I put the paper carefully in my pocket and bought fruit. It was cheaper here than on the ship. Then the whistle blew and we 'fell in' to march outside the building under an immense shed where the trains were being coupled up.

We were waiting for our train, when a girl came along. "Were the 'ell you bloody 'Mericans think you going?" Some one said: "To France, you damned ninny, to whip the Dutch. You British are too cowardly even to try." She thumbed her nose at us and suggested she would be dancing on our graves soon. Then she seemed to change her mind and decided to dance then and there. Some obliging Yank pulled up a four-wheel truck. She mounted it and danced. Just what there was to dance about I failed to see, but she was making a great 'show'. Her physical prowess, however, had not made much of a hit. Probably it was too early in the morning. She might have had better luck after the falling of the shadows. But a 'Sammie' threw her a few *pfennig* (German coin), and she swore the hope that "all you bloody Yanks get killed" . . . and we marched down the platform toward our cars.

Then all of us laughed. A solemn faced Britisher with neatly trimmed beard appeared with a large bundle. He handed to each of us a card gorgeously engraved: "GREETINGS FROM KING GEORGE." . . . Everybody laughed . . . and one boy called out: "George WHO?" . . . The Britisher answered: "His Majesty." And another asked: "Where is George? Why didn't he come to see us? We came to see him." I thought I would keep my 'Greeting' for a souvenir and put it in my pocket but it stuck out and rubbed my arm . . . I threw it . . . It sailed away across the platform . . . Now the air was filled with thousands of flying 'Greetings' . . . "To hell with King George . . . Hurrah for Woodrow Wilson . . . No, boys, King George is all right but he just doesn't like to get up early!" And someone was singing:

"He hates to get up in the morning,
In bed he'd rather stay,
That's the way with the British,
Yip! Yip! Hip! Hip! . . . hurray!"

The fruit-stands in the depot had been a pleasant sight. Edibles had been in demand. But the soldiers were sore at the refusal of the British to accept our money. When at last, however, we did get aboard the train oranges, apples, candy, everything, came with us,—a reprisal against the British, harking back to the days on the EL PENOR.

When once in the cars the men quieted down. They could rest now . . . and eat. They had oranges, but not enough to go around, but soldiers divide and everybody had a bite.

The railroad cars into which we had loaded were o different construction from those of American manufacture. Each car had three compartments. When not in troop-use these compartments were to be occupied by different classes,—first, second and third. In troop-use, however, these compartments accommodated,—and very comfortably, too,—a squad of soldiers. I secured a seat on the right of the train with back toward the front partition. Into the seat, of course, I had to take my pack but this was easy compared with the previous lack of space on the boat. I was now riding in 'solid' comfort.

The railroad system, judged by the opportunities I had to observe, is a good one. Where it was at all feasible the lines were double-tracked. In many places the right-of-way carried four tracks, certain lines being used exclusively for traffic bound only in one direction, which of course, greatly facilitated the movement of freight and passengers, and was a boon in days of war. The cars, according to the United States standard, were jokes, relics of an English aristocracy. The freight cars were small, measuring about seventeen feet long and capable of conveying, according to their markings, two tons. They had a coupling device similar to the old American 'link and pin' used before the days of automatic coupling by impact. On the outer ends of each car was a spring bumper designed to lessen the jolts of contact in motion,—a contrivance American railway executives could well afford to study. To anyone who had seen the immense American system of transportation in operation these trains gave the impression of being toys. The speed of the English trains was terrific, the roadbeds being good and the rolling stock light. The English, as well as the French, we learned later, had built their railroads with the idea of avoiding grades, and as the country in many sections is rolling, they burrowed through mountains and hills in an immense system of tunneling. As we went darting over the country we passed from one tunnel into another, and as there was an inferior grade of coal being burned at that time the pleasure derived from viewing the scenery varied from being constantly marred by a cinder in the eye to complete obliteration in a smoke filled tunnel. All bridges of size seemed to be of stone, the arches of which were real works of art.

After leaving Liverpool we rode in the direction of Winchester. On this day we were served a morsel of English aristocracy, the touch of which flared in the hearts of all Americans. Its putrescence rankles even yet in the minds of the men of the 90th. As we had not had breakfast and the mess-sergeant, of course, had not had an opportunity to furnish food en route, our spirits went high in anticipation when we saw a little city in the distance and the word came back we might detrain and remain twenty minutes. The train rolled to a stop. It was emptied with a dash, and every man with a shilling instinctively sped toward a sign which suggested the willingness of its owner to dispense food to hungry men. Imagine, however, our consuming indignation when we were accosted by a British officer who refused us admittance, explaining that "even at this time Hinglish officers are partaking of their noonday repast within." . . . "Hinglish custom", he avowed, "forbids a common soldier from eating within the same 'ouse with an officer of rank." I had, however, gained entrance to the cafe, heeding not the carefully chosen words of this 'man of rank'. I might just as well, however, have tarried for the Britisher, aided by American troop officers, succeeded in enforcing the custom by my ejection. In condescension, however, the English officer "in deference to the American command" arranged that we might have a cup of lukewarm "tay" (tea) whilst Britishers in their grandiloquence sat at ease inside the cafe.

At this station one of our men was left behind. It was some clays afterwards before he straggled into the company again . . . but he reported having had a good time.

On this trip we went through the heart of the Island passing a part of Oxford University and many other historic places, finally detraining at Winchester. Our trip on this fast passenger train, although fairly crowded, had not been an unpleasant one, as days then went. There was a great sameness about the appearance of the country. The land was in a high state of cultivation. Those people understood the value of intensive farming. The dwellings were all of a pattern,—two stories, box, red brick, covered with red tile, and a protruding chimney. Here was a similarity of construction calculated to lead many a late homeward-bound convivialist into the wrong abode.

We arrived at Winchester some time near the middle of the afternoon. 'Just to get somewhere' was comforting to us, but little did we surmise what was immediately ahead. Winchester was not a great distance from the port of Southampton, in the same province or political subdivision. After detraining, we started on a 'hike' the like of which I had never before experienced. Only that morning we had unloaded from a thirteen day voyage. No opportunity had been given us to "work off our sea-legs", and weakened both from the lack of exercise and the sprained ankle received by the fall in the early morning in Liverpool, I, not unlike the entire company, was in no condition to make the trudge of three miles . . . all up hill . . . to the camp, Winneldown. But off we went. We had eaten nothing except a bite off a divided orange and a cup of bitter 'tay', but we must get to camp. Some of the very best men in the company 'fell out' that afternoon and had to be hauled into camp. Scott, one of the strongest men in the company, went down on the roadside, no longer able to raise a foot. Here it was that Lieutenant John R. House showed himself to be a real man, showing a spirit too seldom evidenced by some officers. The lieutenant came to the fallen Scott who by now had attracted the attention of some other officers. They were standing over him castigating him for his physical weakness. Instead of pursuing the same method, House released Scott's pack from his back, lifted him to his feet, gave him a drink from his canteen, and then the two walked arm in arm up the hill, the lieutenant carrying the pack,—and Scott finished the trip along with the company. Up to this time I had not had occasion to see much of the methods of this officer but I knew then and there he had the 'makin's', as the boys would have expressed it. He was the kind who could lead men, the kind in whom men could place a confidence, knowing it would not be betrayed. In this opinion I was not mistaken.

We arrived at Winneldown before dark, and just before swinging into our company street I chanced to see a British military mannerism which amused me greatly. Captain Whitaker was walking in the fore and to the left of the company when we met a British soldier face-to-face. He cocked his head back in surprise, came to 'attention', and bringing his left hand to the brim of his hat, thumb sticking straight out, held the salute until it was answered by the captain. The salute had much of the appearance of the familiar American hand-signal of personal disrespect. Tired as I was, I laughed out loud, much to the disgust of the commander. This salute,—coming from the hand next to the officer, regardless of the side on which the soldier passed,—was a regulation English salute, however, to which we soon became accustomed.

Another 'square', according to the British vernaculain measuring distances, . . . and we arrived. Before us stood crudely built wooden 'billets',—the English term for 'barrack'. We found our bunks were to consist of three ten inch flexible planks which rested on the floor at the 'foot' end. They were elevated three or four inches by a cross-member at the head. Being only an inch thick these boards had quite a spring in them. They did not, however, have all the comforts of Sealy mattresses, but they constituted a bed of a 'sort', a great improvement over the bunks in the hold of the ship.

The Fourth of July was just ahead of us and a great clamor went up from the men for permission to go to London. The boys seemed to think it quite appropriate that American armed troops should celebrate the day in the British capital. It was at first stated we were to be in Winneldown for a long period of training and permission was refused us to leave the area. Suddenly this order was countermanded and passes were issued. I was still suffering from the sprained ankle, and decided 1 would not make the trip to London on the holiday. I wanted to see London, but I resolved the doubt in favor of the attempt to care for myself . . . but as it turned out I might as well have gone!

On the morning of the Fourth of July after the departure of a large number of the men of the company for London, the stay-at-camp remnant was surprised to be called out for company formation. Knowing this was a holiday and that the British soldiers in the camp also were doing nothing, I was surprised at the order. Imagine my disgust when we were walked about a mile across the dust covered camp and made to stand almost all the morning listening to LORD MAYOR SOMEBODY, His Excellency, The Mayor of Winchester, make an interminable speech about the Fourth of July and "why it had made the British and the Americans such good friends" . . . and the humor of the nonsensical piffle has never left me. What fools even the so-called 'big men' of nations will make of themselves when they do things with ulterior motives! When the mayor finally concluded his speech, not a word of which I now recall, we were dismissed and issued passes to Winchester 'for the period permitted by English regulations'. This, we learned, meant 11:00, p. m.

Winchester is one of the oldest cities in England. It is located in a valley at the foot of some high hills on the summit of one of which was located Winneldown. The roads leading from the camp into the city were superb, although not paved. The hills overhanging the city were covered with a luxuriant green giving the underlying city an atmosphere of quietude and restfulness. Seldom have I ever seen a more luxuriant foliage than was here for the eyes of water-worn soldiers. The trees were majestic, immense, spreading. The ground was free from disagreeable briars and brambles, but interspersed between the hoary old trees, boasting of accumulated centuries, stood a new tree-growth which added a touch of the healthful and virile to the landscape. And when I viewed the city from the hill-top, so quiet was the place, I thought the Englishman must have been looking from my vantage point when he wrote:

"Hush'd are the winds, and still
the evening gloom,

Not e'en a zephr wanders through
the grove."

But when I had secured a pass and entered into the city the British illusion soon faded.

Armed with a company commander's pass, "subject to the period permitted by English regulations" I limped a 'backtrack' across the camp, taking the route only so recently negotiated. To the ordinary person it would appear that an exit from a camp would not have been attended with any difficulties but this was far from true. The camp was large; there were thousands of tents forming streets, and hundreds of wooden buildings gave the appearance of sameness. Directions were hard to keep, but I found the main road covered with a multitude of moving trucks, and knowing the general direction of the city, having previously glimpsed it from my vantage point, I turned toward the city. To my surprise I came upon Howard Edmiston who had 'greeted' me on the day of my arrival in Camp Travis. Howard was disgusted. He did not have a pass and the British guard had refused to let him pass the gate. Intent upon seeing the city with him I suggested he 'run the guard' while I engaged "his majesty's subject" in conversation. As this appeared to be a good idea we returned together to the 'big gate', where we found a line of passing soldiers. The guard was inspecting with meticulous care the pass of each man, and I began engaging him in an undesired conversation. Howard was searching his pockets for a pass. Strange . . . he found one . . . The guard read: "Pass Private Chris Emmett . . . L-Co. 359 Inf. . . . Subject to . . ." The guard accepted it and Howard passed. The lines continued to flow . . . some in . . . some out. Now I could see Howard sitting on the roadside . . . leisurely smoking a cigarette . . . He hailed an inbound American . . . They laughed . . . Howard slipped him something . . . and I talked with the guard: "Beautiful country this of yours!" and it was "a bloody good country" but "You Yanks awsk too many bloody questions" . . . I knew, but I must have time . . . "What is there to be seen in Winchester?" . . . "Have you not been there? O, that is different then . . . Don't fail to see our stupendous cathedral . . . a work of art . . . ages old . . . mellowed by the ravages of time . . . A chawnce of a lifetime . . . Before the war . . . the damn bloody war . . . people even from your country came all the way to view it from the hilltop" . . . An incoming soldier . . . He presents a pass and enters . . . "Hello, Emmett. Going to the city? (I do not know him). Give me a cigarette". I hand him the 'makins'. He rolls one and passes back the bag. I felt a paper and put it in my pocket . . . "What did you find under the hill . . . cafes, whiskey, women?" "Hell no! Cathedrals . . . beautiful cathedrals . . . the home of . . . who the hell was it? . . . a graveyard in the back . . . the sacred last resting place of lord . . . Now who was the lord . . . Who was that lord . . . Ha! Ha! Had to come all the way to England to learn the lord was dead . . . and he's buried back of the cathedral . . . A heluva country, this!" . . . and he was gone, swinging down the road . . . laughing!

Again in possession of my pass I said, "Well, guess I'd better be seeing the works of art, too . . . Seems to be nothing like it" . . . Howard was waiting on the roadside and we went arm in arm. At the brow of the hill we stopped. The city lay at our feet in its tranquil restfulness. I would rest my ankle, but Howard grumbled: "Just my luck to come off with a cripple . . . and the saloons will soon be closed" . . . And that was a matter of concern to me too, so I limped on . . . down the hill . . . Not such hard walking now . . . no packs . . . something to eat . . . perhaps an invigorator down the hill.

We met an Englishman . . . with many whiskers . . . a top hat . . . "O, hell, will these Hinglishmen never learn the war is going on"? . . . More top hats . . . and we were accosted by one . . . "Going to church, my men?" . . . "Church, hell! Goin' to war . . . Where the hell you think you're goin' . . . with a lid like that?" . . . "Pardon me, gentlemen . . . I would not intrude . . . but I observe you are of the new American army . . . newly come!" . . . "O, but he has good eyesight, eh, fellows?" . . . "Your men are taking great pleasure in viewing our great cathedral . . . And we have services there quite shortly . . . Perhaps I may direct you to your worship . . . I would be pleased to accommodate our kind American visitors" . . . "Thanks, if that's the cathedral we hear so much about, let's be dragging it . . ."

And we approached an architectural wonder. Unversed in the history of its environs I could not appreciate the importance of the story our guide insisted upon telling us,—the construction of the building, the sanctuary of the dead, the flowers in the garden imported by Lord So and So from parts of the world unheard of by me. But as uninteresting as was the courteous Englishman's narrative the whole setting of the place commanded the attention of even the casual observer.

The elder English were taking their history seriously. They were proud of their nation's accomplishments. They walked across the floors of their famed buildings with muffled tread. I listened inattentively, but drank in the vision of the great spectacle. Here was a building, according to the story of the guide, the construction of which began twelve hundred years ago. (I only quote him.)[2] As the building was approached it could be seen rearing itself majestically above the surroundings. Located at the rear of a great plaza, or park, it stood behind a colorful garden,—immense trees stretching their boughs with grace and dignity, giving the appearance of an over-spreading deep green canopy flung out to protect a variegated panorama of blossoms. The building, stone, browned by countless seasons and mellowed by age, stood out about one hundred feet above its environs. Our entrance into the building was silenced by the deafening hush pervading the spacious auditorium spread before us,—a space of almost a city block,—all under one roof supported by the walls, counter supported only by a row of columns set back some fifteen feet from the walls,—an architectural feat grasping the imagination! Peering across the immense room, the sun, now fast slipping low in the west, reflected its rays through the most delicately tinted windows. Statues, the handiwork of the adept, relics of an historic past, softened the expansive walls. I could not appreciate the technique of these artists

2 For a most interesting account of this cathedral see CHAMBERS ENCYCLOPEDIA, Vol. X, p. 644 et seq: "Winchester, . . . situated 60 miles WSW of London. It originated . . . in a tribal settlement placed for safety on the summit of a hill . . . As the settlers became more numerous they descended the slope and here rose Wenta, to be known as Wintancaestre . . . The first Christian Church in Britain was built here in 169 A. D. The church was converted into a 'temple of Dagcro' . . . 495 A. D. . . . The fantastic cathedral of the Saxons did not accord with Norman ideas, and Bishop Wakelin . . . demolished it, and build (1079-93) a dark and ponderous pile . . . replaced the small round-headed clerestory windows with large pointed lights, and added an arched stone-groined roof, producing by this transformation the finest Perpendicular nave extant. In the centre of the Choir stands an ancient tomb . . . We also notice those of . . . Jane Austen and Isaak Walton . . . The cathedral is the longest in Europe (556 feet)."

but they impressed me as silent onlookers of a romantic past . . . keepers of the faith!

Back of the main-hall . . . cloistral anterooms, inartistic, almost adobe in appearance . . . "Here, my good men, died Monk" So and So . . . Unknown to me!

Faintly . . . appealingly . . . tugging at my very heart-strings . . . came low, mournful music, a choir singing softly . . . men kneeling in prayer, dressed in black . . . women audibly sobbing . . . "Now, let's go into the cemetery,—the burial place of our renowned dead, where . . ." I did not care 'where'. Thanking our courteous guide, I walked, casting a glance into the cemetery, stopping only when I had reached the bright sun-light and a convenient park-bench. The sun-light was a relief. The music had gripped me. The requiem reverberated in my ears and I was depressed. I could understand now: it was not the cathedral alone of which the British were proud. They were proud of their past: they were grieved over their plight. They were turning to their religion for solace . . . and here was no place for a soldier!

Howard and the other boys came back to me. They were impatient, inquiring where a saloon could be found. "Saloons? . . . Saloons? . . . And just what is that? O! perchawnce you inquire of our 'pubs'?" . . . They were closed until six, "our eating-hour, you know, old tops".

Here came soldiers,—Australians, fine looking men, grand physiques,—striding nimbly with the appearance of American western-men. They hailed us, and fine fellows they proved to be. They would show us the city . . . Did we have money? . . . They would divide. "What's money anyway? Glad to divide . . . May not see you chaps again." . . . We rode a lorry and they insisted upon paying the fare . . . and 'would we take a drink?' . . . We would,—and they knew an accessible dispensary where they could gain admittance. They knew how . . . A drink . . . and we felt better. Then a long trudge around the city. There was nothing to command an interest,—just a city filled with men using walking-canes. We laughed at this but the Australians reminded us it was a 'craze come out of the war' . . . "You will be using them soon . . . A great aid to walking with a heavy pack . . . Not a bad idea you will learn . . . Flag-flapping imitators, we admit,—but a great aid to soldiers on the march."

Heading down the street we came upon a crowd of men from our camp, many of whom were my friends. They were going 'across the tracks' . . . "Great times there" . . . but in my ears echoed the cathedral sounds . . . "Holy . . . Holy . . . Holy . . . Lord God Almighty . . . Maker of Heaven and earth . . ." I refused to go, and suddenly deciding I was hungry, made for a small cafe. Here many men clamoured for food. British officers came in, ladies on arm, and departed again. American officers appeared and sat at tables with their men. A British sergeant arrived. He grumbled, "Too many damn Yanks these days". One of our men thumbed his nose at him . . . "You bloody damn Yank!" . . . and there was a scramble. I caught his arm. An American officer intervened. We stood at attention. "All right, boys: eat your dinners . . . No cause for trouble." Then we sang:

"We came all the way to Tipperary;
We came all the way to see,
If a damn lousy Hinglisher,
Can whip any of us three."

Back in the street again it was dark. The 'pubs' were open everywhere. Englishmen sat sipping their ale. American soldiers lolled at the entrances. There was a rush from a side-street. We saw our friends who were going "across the tracks". They were running. There had been a fight . . . some-one hurt, but what was it to us? . . . Just an Englisher . . . "talked too much!"

Another long trudge through the city brought us to the edge of the business district where there were suggestions that we go to camp. Others suggested we drink. The latter prevailed. The streets were noisy with drunken soldiers. We thought we were sober and sought a quiet place, a side-street and an open door, and we entered. The place was small. British officers sat at tables, monocles to eye, sipping. There was a wave of an officer's hand. The bar-keeper came to the door to close it announcing, "We do not serve any more tonight." I said: "The hell you don't! Give me a drink of Irish whiskey!", and threw open the door again. Again the bartender said that the place was closed for business and he started toward the door. I kicked at him, and missing, struck the door which opened up square in his face. He fell heavily to the floor. The British officers ran, and when the bartender arose, bleeding from the nose, with a cut across his forehead and a swelling lip, the house was clear except for American soldiers and the keeper of the 'pub', who proved to be a good sport. Responding to our pounding on the bar . . . "We want Irish whiskey!" . . . he bowed graciously,—a blob of blood dangling on his nose-tip,—and inquired: "Now, my good men, may I have the pleasure of serving you?". We took Irish whiskey . . . green bottle . . . full glasses . . . more whiskey . . . And then someone sang:—

"We won't go home until morning,
We won't go home until morning,
We won't go home until morning,
'Till daylight doth appear:
We're a bunch of jolly good fellows,
A bunch of jolly good fellows,
A bunch of jolly good fellows,
Partakirt of some of your cheer."

And from behind the counter came: "Aye, Aye, Sirs. I'm certain I am agreed."

When we had again become less vocal, the bartender suggested we had drunk sufficiently to the Irish and proposed a 'bottle of Scotch'. His proposition was accepted.

Three drunken soldiers started homeward. There was still the trouble of the 'Big Gate' to come. However, the guard passed us with a "get to bed before some bloody M. P. finds you."

Straggling along the road I decided to rest my ankle. I chose the front of British Headquarters where I sat upon the ground. Then all decided to sing. Howard was disgusted, and slapping his hand over my mouth, said:

"Nobody but a damn fool from Hamilton, Texas, would try to break into the guard-house", and jerking me to my feet, we sauntered off into the darkness. He reached his quarters. I must have arrived also, for the next I knew the soldiers were returning from London. It was still night, and there was much noise. King and Meeks were happy . . . and someone told me the story . . .

The party which had gone to London had not agreed with the British any better than had we. They too had found the 'pubs' . . . many girl bartenders, tough-sisters . . . much swearing . . . more drinking . . . Then King offered to buy a drink. The 'sister' drank with him, but refused his American money until it was exchanged. The proprietor agreed with her. King reminded them of the fact that when England came borrowing from America to whip the Deutsch she was glad to get American money. The argument started . . . It ended as did the War of 1812 . . . with many shattered glasses, but the British capital was not evacuated!

Our mess at Camp Winneldown was excellent. The kitchen was clean and the food good. Plenty to eat and a rest the following morning put us in trim again. But orders came that we should drill as usual. Lieutenant House assumed command of my platoon. Being a man of wisdom he obeyed only the spirit of the order, and we left the camp headed toward Winchester, he saying to us that no soldiers under him would have to work when there were sights to be seen. Imagine our surprise when we were met by a British military squad who refused us admittance to the city. The explanation was given, however, that an American troop had recently shot up the town, and the order that 'No armed American troops are allowed within the limits of the city' was a general one.

Then we lay under the shades of the big trees overlooking the city and slept with arms stacked, . . . a quiet evening, an understanding commander.

Bugles called us from our beds early the next morning and orders were given to 'pack up'. It is surprising how easy the trip back into Winchester was made. Now we were rested. We swung along in the early morning, with the dust rising into our nostrils, but there were no grumblings, no complaints, only songs could be heard along with the crunch, crunch of hobnailed shoes striking the flinty road.

A stop was made near the depot: "Fall out" . . . and we bought apples . . . Then "Fall in", and boarding cars we were whisked again rapidly over the rails. Our train scudded through fair sized cities, but nothing of importance passed before us. And we stopped again. Before us was a great shed, immense ships docked side by side. Boats nosed against the river-bank as far as the eye could see. It was Southampton. Thousands of American soldiers were under cover of the shed. Soldiers, a sea of American uniforms . . . and we had nothing to do but lie on the floor. Tiring of this, I found Rogers Tipton and we walked down the dock to see an arriving vessel discharge its cargo. A few British officers came down the gang-plank. Then there were moans and cries of pain above the din of moving feet, and litters were being lowered from the ship. It was a British hospital ship from the French coast unloading the men from the northern fields of Belgium. God! and what a sight! Bandages, blood soaked uniforms . . . moaning men . . . litters covered with sheets . . . and Tip turned to me: "I can't stand this any longer. I'll be one of them soon enough." . . . The sun fell rapidly in the west as we watched a constant stream of ambulances arriving, taking their loads and noiselessly slipping back into the big city.

"Fall in", and we were marching again. Around the corner of the great shed we could see the steady streams of khaki flowing into sixteen boats resting at the wharf. My boat was near the shed. It carried no guns. We were told we would have no escort, and each man was to provide himself with a life-preserver . . . The lousy things! . . . Filthy! . . . Vermin covered! . . . But we were forced to put them on: "Better cooties than drowning" . . . "Little choice", I thought, as I strapped it on.

Just to the right of my boat was a relic of the war, an immense freighter with a torpedo hole about twenty feet square in her prow tugging placidly at the anchor-chain. She had received a 'Blighty'. She had come home for the "period of the emergency", for no one had time to repair her now.

The power of our boat was turned on. The pulsating screws could be felt throughout the vessel as it tackled the placid current of the river, and soon we had acquired a tremendous speed. The river was narrow and the channel probably deep, for a constant stream of inbound boats and ships, some gigantic in size, were passing us. There were few signs of life aboard. Every few minutes a hydroplane would slip out of a clear sky to dive into the river like a duck completing its flight for the day. Immense castles skirted the water's edge. A vast expanse of blue shimmered faintly out ahead, and the land receded to each side of us and the coast line swung back from the mouth of the river. Spurting into a new frenzy of speed our little boat laden to its full capacity with khaki uniforms soon passed out into the rolling waves of the English Channel. England was behind us. Dusk began to fall around. The water became 'choppy', the boat lurched and churned and bitter cold settled upon us. I had been sitting on the first open deck and, although uncomfortable, I endured an increasing misery attempting to ward it off with, first my overcoat, then my pancho until the cold penetrated my bones and my teeth chattered. I stayed until darkness enveloped us. Then, I went behind one of the smoke-stacks, but the smoke was stifling. I retreated again only to find that all the boat's soldiers were doing the same thing. We tried to sing, but an officer cursed us: "Did we want to be sunk"? Then some one chanted (to the tune of 'We won't go home until morning'):

"They sank us all in the channel
They sank us all in the channel,
They sank us all in the channel,
With the British lying nearby."

From the top-deck, sitting astride a pile of baggage, to the second deck, seeking protection from the cold, I worked my way into the hold. Such another miserable, cruel congestion, I have yet to see! Soldiers, soldiers prone, soldiers trying to rise to their feet, other soldiers stepping here, striking an outstretched hand, hobnails digging into flesh, curses, vile epithets,—a miserable jam of soldiers! Working against the current of moving bodies I found standing-room in the forward end of the hold, and there I sank down. Sergeant Dennis E. Meeks,—that princely fellow of thirteen years with the regular army, a gentleman, an excitable 'scrapper', loyal to his company, afraid of no man,—had been in the hold with me. The stifling stench, odors of unwashed human bodies, drove him toward the fresh air. On the congested steps he met a British officer who informed him he MUST get himself a place to sit and 'stay there'. With deference due to rank, Meeks informed him of the conditions below, of the stagnation, the foul air, the scramble of men. "Damn the conditions! I am in charge of this boat. You cannot come above. Never before have I had trouble handling troops." . . . "Never before, eh? Well here's some for you", and slashing fiercely at him, Meeks struck him to the floor, and walking over him, went his way.

A short stay in the hold of the ship and I became immune to the stench. I began to get sleepy. The boat rocked, rolled, jerked. Sick men vomited . . . Others cried . . . There was always a song, some vile parody . . . I slept . . . and I felt the tremor of the boat lessen . . . and all was quiet. The hold was dark, pitch dark . . . an interminable time later a light began to show, and I was awake. "L Company! L Company! . . . Outside! . . . Double time!" . . . Meeks' piping voice. "L Company;" yes, that was me. . . I scrambled to my feet. Men were marching, and I pushed my way onto deck. The morning sun was shining, but a haze, far out, lay over the water. Only a few vessels were within view. The water was placid. Such a change from the night just passed!

Editor's Notes

The troopship that Emmett and his comrades took to England was not the El Penor as he recalled later, but was actually named the Blue Funnel lines cargo vessel Elpenor. Built in 1917, the ship survived both wars and was scrapped in 1952. Survived is perhaps a overstatement. In November 1942, while carrying supplies during OPERATION TORCH, the Allied invasion of Northwest Africa, the Elpenor was bombed and sunk by Major Joachim Helbig, pilot and Commander of the 1st Training and Demonstration Wing of the German *Luftwaffe*. Raised and repaired, the Elsenor continued serving to the end its second of two world wars.[3]

3 (Ed.) Launched in 1917, the Elpenor made several voyages carrying supplies and troops during World War I for the British and the United States. In 1935, it was transferred to the Glen Line and renamed Glenfilas until 1947, when it reverted to the Blue Funnel line again as the Elpinor. In 1950, it was again obtained by the Glen Line as the Glenfinlas and scrapped in 1952. http://www.theshipslist.com/ships/lines/bluefunnel.shtml Accessed 9 March 2017. Franz Kurowski, *Luftwaffe Aces: German Combat Pilots of World War II* (Mechanicsburg, Pennsylvania: Stackpole Books, 2004), 89; Donald A. Bertke, Gordon Smith, Don Kindell, *World War II Sea War, Vol. 7: The Allies Strike Back*, (Dayton, Ohio: Bertke Publications, 2014), 320.

The Elpenor on August 11, 1918 arriving in Birmingham Harbor with a load of Doughboys headed to the Western Front. Note the dazzle paint scheme meant to confuse German Uboats as to the direction and speed of the steaming ship. U. S. Army Signal Corps photo.

A group of Doughboys on an unidentified troopship headed "Over There." While a number of their curious comrades watch, a KP (kitchen police) detail is "spud bashing" and peeling potatoes rather than participate in a submarine sighting drill with life jackets. Stieghan Collection.

Chapter 6

Training in France

I was eager to get off the boat, and jumping from deck to wharf I lined up with our fast forming company. The city was before us,—Cherborg, France!

My first impression of the city was a pleasing one. Houses, most of which were two stories and of Corinthian architecture, the delicate lines of which shimmered a welcome to us in the morning sun, stood back an impressive distance from the water-front. From a distance it was a city of white houses. And we turned our backs upon the water and marched . . . and then . . . O, then . . . what a revelation. I had thought of France as clean, delicate, and dainty, of her people as those of artistry. But here children ran out into the streets. They were in tatters, unkempt, unclean. The women,—with the ravages of age, bowed of shoulders,—lacked all the sprightliness I had expected. Surely this was the devastation of war?

Streets were paved with stones,—rough cobbles. They hurt our feet, and we staggered and stumbled. Murky, putrid, slow moving streams of water oozed along the gutters of the streets. Children waded in the filth . . . wooden shoes . . . bare-feet . . . A fetid stench assailed my nostrils . . . Old men came to the doors . . . black hats, and capes thrown around their shoulders . . . more stooped shoulders . . . and this was France!

We stopped in the street to permit a column of soldiers to cross ahead of us. And there was a saloon . . . Vin Blanc! I wanted a drink, anything to get this stench out of my nose. Lieutenant House seemed to be reading my mind. He was saying: "It looks enticing, boys, but let's go easy. We'll talk about the French wines someday." . . . "Forward" again, and we struck uphill. It seems the infantry always marches uphill! A few more Frenchmen. They cheer. Others peered at us stoically. No emotion. Did they care whether we had come or not? These drab . . . yes, drab, . . . men . . . perhaps drab women!

But the air was clearing now. The water no longer ran in the gutters and there were fewer houses. The morning breezes were wafted to us, . . . and we were happy. The city was behind us. Trudge . . . trudge . . . trudge, and we could see down into a valley. We met American soldiers: "The 'eats' are fierce." Little did we then know the import of the words, but we had not eaten breakfast, and we could not think of any food tasting bad. Two kilometers (Instinctively we began to measure distances in kilometers.) and we came to a most beautiful park. Great oaks, flowers, tranquility! We wanted to rest, but we turned again uphill. The earth was rugged and broken with the imprints of many wheels. And there was our camp!

The camp was only large enough for two regiments of soldiers. When informed it was a 'British Camp' we had visions of a clean kitchen, good food, of Camp Winneldown,—but the surprise! Located on a hillside the drainage was good but as rain had fallen ahead of our arrival the ground was soft, although not mucky. There were too many small stones for that. We halted before some very low squad tents. I lay down with relief, being informed there would be no more duty that day. What a surprise! I roamed about the camp, found a canteen manned by British men,—the first I had seen in canteens. They were selling many things to eat. We ate sparingly. Our money was now getting low. We drank French wine . . . sour . . . not what we had expected . . . A bugle call and a rush for dinner. What an awakening! And the British fed us: a thick slimy stew, hard French bread, sparingly served, a damned conglomeration tasting of lemon and orange peel . . . "Marmelade" said the British . . . "Fine" . . . "Fine, hell"! and I threw mine away. A rest again and another bugle call, but I had heard the rumor: 'The British were feeding us horse meat', and I wanted to know. I went to the kitchen, a large wooden frame building. I wanted to see. I announced to the guard at the kitchen-door that I was an assigned K. P. He passed me. What a stench? The floor was ankle deep in slime, dirt, filth, scraps from the table, mud from the shoes of passing thousands. A hulky butcher was cutting meat, swinging a heavy cleaver. And a skinned horse,—the whole carcass,—was lying on his table. The knife chopped through the flesh, and a cut,—some ten pounds,—fell to the floor, disappearing below the slime. He grabbed it up and threw it into the stewing kettle. I gagged and retreated from the building. The rumor was true.

Although there was a distinct shortage of food at British Camp No. 1, we suffered little because of the presence of the canteen from which could be bought,—if you had the price, and it was an exorbitant price at that,—many of the American canned products. As our sojourn at this place was marked by no outstanding incidents our two or three days there passed rapidly and rather pleasantly. Just before going away, the order having reached us in the early morning, Sergeant Meeks mounted a truck and told us that "even a one-eyed man can see we have hubbed hell, and it is advisable for us to purchase such food as we can pay for and carry." We obeyed the admonition, much to our pleasure and comfort later on.

Out of the camp we went again, back the same road. We were happy that British Camp No. 1 was being left behind us. What did we care where we were going? We were going: that was sufficient. Rumors! Rumors! Rumors! The air was full of them. "Going to Italy" . . . "That's all right by me" . . . "Headed for Belgium" . . . "All right. We won't have to eat horse-meat anyway, if we can get away from these British" . . . "Going to Paris" . . . "O, boy! Not so bad, eh?"

Swinging along the road we met a wagon on which was loaded two dead horses. King was shouting: "O, you damn Limeys: We beat you to it. We are leaving. Eat 'em yourselves!"

Again at the top of the hill, we could see the placid channel, a pinkish haze lying over the water. A long neck of land stretched out from the city. Small boats,—some sails,— . . . What a beautiful picture . . . but the reality back of it!

We were herded into a barbed-wire enclosure, a sort of stockade. There was no shelter and the damp biting wind blew off the water. We huddled up close to each other, lying as near the barbed-enclosure as possible. We waited for hours, and I thought of the cattle-pens of Texas, cattle lying close together for -protection against the elements. So often had I seen cattle drift to the fences and lie down.

They served us 'hard-tack', a tooth-breaking substitute for bread. And the British served us 'tay', but it was warm and we drank it thankfully, for the mizzle had dampened us and the cold breeze off the channel seemed to drive the cold deep into our bones. We lay closer together and waited, waited, waited. Would we never move from there?

The sky smudged over with a smoky pallor and the evening drew toward a close. The gate was thrown open and we arose to our feet. Our joints were stiff. We were cold, sulky, and hungry, but we shambled out to watch a 'frog' engine feebly switch its tiny cars, which, after an interminable time, were shunted alongside us, and we loaded in. The cars stood a long time and we took time-about looking out the side-door. French women, French children, crippled French men, walked along the side of the train with palms upturned . . . begging, always begging. Would we never find people who would not beg? Had the war made a beggar nation of France?

Our train stood on a track which paralleled a street of closely built houses. In these were billeted French colonial negro troops. Our American negroes were perfect blonds by comparison. And just a little farther down the street was a troop of Afghanistans with faces hard and symbolic of crime. And when I assimilated our surroundings, I realized that, had our stay been prolonged, there would be inevitable conflicts between our boys and these people. I was glad we were loading out for somewhere else.

A passing American negro met a French colonial negro, and addressed him in English, asking, "How long you done bin over here, Big Boy?" The Colonial replied in a tongue unknown to the American negro, who answered in disgust, "Huh! you done bin heah so long youse forgot yore own langwige."

A man came from one of the houses and stood below the door of my car. He was trying to tell me something, but I could not understand him. He was white, but he did not understand English, and that, at the time, impressed me also as strange. I, too, had always thought in terms that all white men spoke English. A child of a nation of self-sufficiency! But this man wanted bread. Someone understood him and interpreted for us. He had been in the army, and had been burned by mustard gas. The skin of his face was a seared scar. One hand had been burned, infection had resulted, and amputation at the wrist became necessary. The other hand had drawn until he could no longer make use of the motion of the fingers. A physical wreck . . . yet alive! What a pity!

It required only a glance for us to comprehend the significance of our means of transportation. It was the long-heard-of 'Frog Train' with "side-door Pullmans", the cars of "Forty and Eight" fame, never to be forgotten by the A. E. F. Troop-cars in fact were nothing more than the ordinary French freight cars transformed to carry men and animals. The length of the cars was about seventeen feet with a width approximating that of the American car and a height of about seven feet. On the inside and at the ends a bench was provided. There were no other conveniences for either man nor beast; however emblazoned on the side of each car was printed:

"Cheveaux 8 {en longe)
Homines 40."

In other words, these cars would hold eight horses, (if placed lengthwise,) or forty men. Since we never found a comfortable position in the cars we concluded our loading also should have been 'en longe'.

And just as the sun faded out in the west our long train with forty heads peeping from each door pulled slowly southward toward our destination . . . the unknown!

There is no reason why the common soldier should know his destination and there is every reason why he should not know it, but the fact that soldiers were not informed always caused rumors and discontent. We could tell by the position of the sun after our first night out that we continued to keep a southward course.

Instead of a 'bon voyage' the trip was concentrated misery. Lacking space we took turns at lying on the floor, trying to get a little sleep while the other fellow stood up. The cold night-air swept through the cracks of the car and we shivered. A big-footed soldier, enduring his misery as long as possible, would decide he had to walk, and out of his 'place' he would come taking no small portions of the skin from the hands of the prostrate sleepers with his hobnailed shoes. The meanderings of these somnambulists would invariably call forth an argument,—perhaps a fight.

It is a rule amongst soldiers that those who really wish to fight shall have no interference from any quarter. Fighters are permitted to satisfy their warish lust unmolested. We had a big red-faced fellow named Crow, hailing from Henderson, Texas, who had grown up to respect the rights of others and be certain no one was disrespectful to him, and coming from Texas, it is an almost sure bet he felt that an instrusion 'from out of Mexico' was wilful. By chance a Mexican boy, named Florez, from San Antonio, 'absorbed' Crow's place while 'Old Red' was leg-stretching. His refusal to evacuate caused Crow to evidence an immediate determination to cast the 'pepper-pod' from the train while the train was running at high speed. Knowing there had previously been bad blood between the two I knew Crow actually intended to carry out his threat unless interfered with. Crow shoved him toward the door whilst the Mexican begged for his life, and when some of us interfered, the Mexican contracted his corpulency into an infinitesimally small sector far back in the corner of the car where he contented himself the remainder of the trip.

The ride through France was an interesting one, although we were not at all times informed as to our location. A stop at a station, however, afforded me the opportunity to buy a map from a 'Frog' and our car of boys whiled the time away guessing at our final destination. Glimpsing the names on the stations and checking on the map, we soon found our general course. We passed through some of France's most picturesque territory. The train, running down through the valleys, high hills to either side, vine-clad gardens terraced to the crest of the heights, castles,—evidences of an age-old civilization,—gave us food for thought.

Of course, stops were made now and then. At the rail-stations the French were importuning us to buy. We bought, and the cars were loaded with wines,—vin-rouge, vin-blanc, vin-ordinar, champagne, cognac,—

and old King said: "Now we can all get drunked-up."

By way of parenthesis, for fear that the ever reoccurring question of the use of intoxicants may seem to be over emphasized and create the impression that there was a real drink-problem connected with the use of the grape-products by soldiers of the A. E. F., I wish to state that there was not in fact an extensive over-indulgence by American soldiers. This may be explained in this manner: There was little whiskey drunk; wine was the chief intoxicant in France. Of course, there was a considerable quantity of champagne consumed, but drinking was ordinarily done when soldiers were off duty or on leave and not at such times as intoxication (which of course often happened) interfered with duty. To this, of course, there were many exceptions, but we also had many, many men over there!

The use of the lighter wines was, to my mind, at times, a distinct advantage to the ordinary soldier, especially under some conditions. It is to be kept in mind that these men, new in the business of being soldiers, were under a most terrific mental and physical strain. A semi- intoxication acted as a relaxing agent, a 'forgetter', a nerve-quieter, and a means of bringing men into a comradeship, resulting in solidification of organization pride.

As to the use of cognac and champagne, under ordinary conditions, these were not indulged in, for the simple reason that most men realized their deleterious effects and chose the lighter wines as a consequence. Then, also, there was an army regulation proscribing the purchase of the stronger vintages, but the order was not as effective as was the common sense of the men. "Off Limits" were placed everywhere by army authorities, but by ingenuity the soldier evaded these regulatory measures at will, in which evasions, of course, the profit-taking French connived.

Our journey took us through the cities of St. Lo., Vire, Domfront, Alencon, Le Mans, Angers, Tours, Bourges, Nevers, Charolles. From Charolles we turned northward, passing through Chalon-s-saone, Dijon, and stopped at Gray. This latter town is called Recy-sur-Orc by the French. The Americans, however, deemed it wise to rename the place which met the approval of all passing members of the A. E. F. who had many occasions to call the name during our sojourn in France.

I noticed some very interesting things about Tours, which was a fairly large French city. One point of interest was the railroad yard, which was in a very deep cut, or excavation, which I judged to have been nearly one hundred feet deep. The breadth of the rail yard was not great, but it was of sufficient breadth to accommodate a great volume of business. To have found a rail-yard "made in the ground", so to speak, was not so surprising as was the care given to the side of this great slash in the earth. The pit sloped gently upward where it was crowned with a variety of flowers, interspersed with gently drooping trees, and draping down the wall-sides were vines,—grapes,—verdurous grapes at this season of the year,—the pride of all France. What a sight! . . . but they were not ripe.

Here we had reached the headquarters of General Pershing. Chaumont, however, became the general's center of activities later on during the course of the war. As we passed along we recognized names now growing familiar to us,—Tours,—the headquarters for the Service of Supply,—and (at the time we were there) the General Post Office of the A. E. F. This distinction, however, was lost in favor of Bourges. As we passed through Bourges we were impressed with the great building activity, a French city taking on much of the appearance of an American cantonment city . . . and high over one of the buildings 'Old Glory' was flapping gently in foreign breezes . . . And our hearts were filled with joy. Something American again!

While we rested for a short period, 'know-it-all-Americans', who had preceded us to Bourges, came to our cars to advise us they had 'heard the news' . . . and "they had it straight" . . . "from a colonel who passed through here last night",—"You boys are going straight to Italy" . . . I looked at my map and formed the same opinion.

We pulled on to Nevers and here saw a wagon-train, American troops, headed toward the western front. Cannon . . . cannon . . . cannon . . . long lines of cannon clucking over the cobbles of the city . . . drawn by sleek, sturdy mules . . . and this looked like Texas.

But at Charolles the highly prized information of the 'know-it-all' faded into nothing, for our train changed its course to a general northeasterly one, and we were left actually in 'darkness' when we stopped for the night on a prairie. Night came and we did not move. I crawled from the car and slept on the ground,—a great relief from the crowded car. When morning came, we were served coffee, hot, piping hot, and I felt good.

Ahead of us was Dijon which showed to be a fine city. When outside the city a few kilometers we stopped alongside a shell factory. My car stood where I could see the process of the manufacture, and row after row of big shells were being stacked alongside the tracks . . . I began to think probably this war was not going to be so rosy after all. Just beyond the stop at the shell factory our little 'pop-corn' roaster (as the boys called the French engine) whistled violently and 'stuck' on a grade. A big United States engine with majestically rhythmic strides came out to 'her boys' and pushed us over the hill and we rolled down into the engine-yards.

Here we saw all species of engines, every make, and from every nation, it seemed. And by the side of our 'pop-corn-popper' chugged an American mogul . . . and there was satisfaction there . . . America was making her engine-power to be felt . . . and I thought there was a parallel to be drawn . . . when the T. O. struck its blow!

We slid out of the yards, into Dijon, past the station and came to a stop in a siding. We were told we were to eat. Just say 'eat' to a soldier and you please him. There was a Red Cross station and women here, and we formed lines, and discarded our recollections of British Camp No. 1, eagerly stepping toward a tendered sandwich and a cup of coffee. Hardly had we begun to 'line-up for chow' until the grounds began to fill with children,—thousands of children with sad sunken eyes, little ragged tots with extended hands begging a crumb, just a wee piece of bread,—hungry, emaciated bare-bodied wretches,—the product of war. With tears in American eyes the boys from Texas and Oklahoma patted a tousled head while the little urchin partook of the last morsel . . . We climbed back into our cars and were silent. We had forgotten our hunger as we looked back upon a stark reality.

Northeastward we continued until we came to an American camp. According to the orders we were to have a night of rest. No accommodations were provided, however, and I rolled out of the car to sleep upon the ground where I could stretch my legs. Just what could be more miserable than a Side-door Pullman, a Cheveaux Special? But a Frenchman with a twist of his shoulders and upturned hands answered: "par bon." Evidently I had not learned all yet.

Breakfast call came again. Lieutenant Montgomery Fly came to me commanding that I take another man and report to Colonel Cavanaugh who was issuing two G. I. (galvanized iron) cans of coffee for each car. Joe G. Lloyd, corporal in charge of the car, at whom I winked, understood there was a purpose when I left with six men instead of two. We picked up three of the colonel's cans of coffee and were away into the milling crowd before he discovered his loss. Lieutenant Fly followed back to our car and asked if I received our quota of coffee. I replied: "Sir: at least that much" . . . "And I thought you had!" he said, as he passed along.

While unloading at Recy-sur-Orc I stepped on a stone which again lamed my ankle, and rather seriously this time. I suffered from the injury for a month, but as I was intent upon 'carrying on' I did not lose a day from duty. My suffering might have been alleviated considerably except for the reputation our medical officer had acquired. The boys said of him: "He is either drunk, drinking, or will paint you with iodine and mark you duty." I, therefore, concluded his services would be of no benefit to me, and with the daily aid of a buddy, bandaged my own ankle and kept going.

The infantry always climbs a hill, so as soon as we detrained at Recy-sur-Orc we pitched our 'pup' tents back from the brow of a nearby hill. A gurgling brook flowed at our feet and wound through the valley with its ripples shimmering a welcome to us. Bill Goodson and I accepted the invitation and took the first swim since we had entered the army in water so cold that a contemplated pleasure terminated in rigors. We thought we needed a stimulant after the bath so we took the road back through the little French town where we were met by an 'Off Limits' sign . . . and an M. P. who believed it to be his duty to enforce the regulation. We also met a Frenchman, who had learned to count American coins,—and he let us in at a back-door,—and when we went back to our tents the rocks under our blankets did not present the previous obstacles to peaceful slumber.

The next morning we had another of those much detested inspections, and we found we had lost much of our equipment in the melee. Captain Whitaker offered to make me his liaison corporal, but I preferred to stay in the ranks. He appointed Almy.

Out on the road again our column headed toward a rising sun. Trudging, trudging, with tired bodies and stinging feet, we swung along the cobbled roads . . . Would we never get there! We did . . . and it was Colmiers La Haut.

I rummaged around in a Frenchman's loft and found a couch, archaic with age and covered with dust, but it was soft. Joe Lloyd found another, but when we had lapsed into peaceful slumber an officer appeared. He ordered that the couches be taken to his quarters . . . a power he had, but a right he lacked . . . but Joe and I slept on the cold stone floor.

Shortly after arriving at Colmiers La Haut, I was transferred to the fourth platoon and joy was mine when I found my commander was to be Lieutenant John R. House. I was made a corporal and assigned to the seventh squad where I met Clarence Oliver. We were in accord from the very moment we met. He was all a man could expect of another, quiet, honorable, intelligent, sober. I do not recall having heard him use a word which would have been distasteful in the hearing of a lady, and, with all, he had a geniality of spirit which attracted him to all men. With him I 'bunked' all the time we were in Colmiers La Haut. He told me very little of his personal history. He was always cheerful and ready to help bear the burden of any soldier.

At Colmiers La Haut we were confronted with the first necessity always to be found in every French town,—that of cleaning up the town. It was a small village of some two hundred population, but there was more filth and dirt to the person than in any place ever seen by me. The cows and horses trailed through the streets of the place, early in the morning and late in the evening, distributing their dung as they walked, and before we could drill upon the streets of the town we became scavengers en mass. Back of each residence was a dung pile carefully thrown high by the frugal hands of the Frenchmen, a reeking, stinking, stench-pot, crying out to

High Heaven! What a smell! And the French objected when we removed it. Their wealth was measured by the height of their pile!

The town was furnished with water from a gravity pump, but as the water-supply was limited and the demands were augmented by our presence the pump seldom worked. Consequently we carried our water nearly one half mile from a spring. The nearest bath was a natural pool some two miles away. In the village a curious sight was always to be seen. Concrete pools in the center of the town had been provided for the washerwomen . . . and all French women wash, it seemed . . . With the rising of the sun, weary women and little girls could be seen kneeling at the edge of these pools . . . beating . . . beating . . . incessantly beating . . . their clothes. And "we learned about laundry from them."

Now work began in real earnest. Drill . . . drill . . . drill . . . and our feet ached. Bugle-calls came early, and we were washing our faces at the town pump. Breakfast, a hasty police-up of the town, and 'fall in' again, and off to the drill-field nearly two miles away, came each day. French officers came to us and we were instructed in extended formation. We ran down the hills and up again with seldom a breathing spell, and the men began to grumble. The food was good . . . but it seemed the quantity became less and less each meal. We knew of no reason for this. Once in a while we would have occasion to eat at the messes of other companies. Their messes (so the boys would bring back the story) were good, and there was no shortage of food at any mess except ours, and the men became excitable in their imagined, if not real, weakness.

A traveling canteen came to town and left some of its wares, consisting principally of delicate foods, at the post commands in the town. This information, magnified in its telling, came to the ears of the men, and the grumblings of the soldiers took on a deep rebellious tone. Bill Goodson was on guard one day and his duties gave him access to one of the officer's quarters . . . He divided the spoils with me that night, and those who heard we had stolen complained bitterly to us and against the officers in the town. They insisted upon having as much as we had. Then, when a meagre meal was served after a hard day on the field, almost by one accord, a mob of hungry men assembled to consider the organization of a band to ransack the kitchen. Sergeant Meeks was told of the action of the men. He hastened to us as we marched toward the kitchen. He mounted an old rock fence. Pitching his voice above the clamour of the half-famished men, he commanded a halt. Some failed to do so, but the general respect for Meeks alone made all but a few obey his command. He then told us in the most positive terms that he knew the cause of our discontent; that a raid on the kitchen would be mutiny; . . . and mutiny meant death; . . . that we were his friends; but as sure as he was commanded to stop the riot he would stop it at any cost. This brought the men to an equilibrium but it failed to appease their wrath.

We drilled the following day with venomous tread. As the company returned toward the town at the close of the day old Bill had a happy thought (which he put to music to the tune of "We've been working on the rail- road"), and he 'sounded off with:

"We've been drilling on the hillside,
All the live-long day,
We've been drilling on the hillside,
While waiting for our pay,
Don't you hear the captain shouting,
"Automatics Right"!
We could do a damn sight better,
If he'd keep our bellies tight."

A ripple of laughter went down the company. The tension was relieved. Tired men washed their faces at the trickle from the town-pumps, all the time improvising additional verses of "Automatics Right", and forgot their troubles.

Each Friday we went on a seven mile hike. Some of these maneuvers stretched themselves out into many more miles. Over hills, climbing, always climbing, our hobnails cutting into the hillsides, scrambling along with a weight on our backs which bore down upon us, we thought we could not pick up another foot. With currents of pain coursing up and down our backs, we carried our packs hour after hour so we could, as the officers told us, "get used to it". Each additional pound on one of these hikes meant much to a weary soldier.

Sergeant Nash, a diminutive man, would always come into our billets fresh and frisky. An officer must have had a surmise, for Nash was confounded beyond explanation one evening when the company was lined up at the completion of the hike and the order given: "Unroll packs for inspection." And Nash was caught . . . not with a blanket-roll and full equipment . . . but with a light blanket rolled neatly around a stove-pipe!

The company had scarcely arrived at Colmiers La Haut before the men of the companies billeted there began to take the feminine census of the village. Our company adopted a little French girl about seven years of age and on her was showered every kindness and delicacy possessed by the men of the organization, and when we came to march away . . . but that is another story yet to be told!

Down in our part of the village there resided a very talkative old woman, who was stooped and grey, but years had not diminished her aggressiveness, even though age, supplemented with constant consumption of sour wine, had reduced her black-stained teeth to mere snags. She expressed an eagerness to learn the English language and appeared to be immensely impressed by the uniforms and insignia of the officers. Some of us decided to give her instructions in English. Therefore, with the little French which was rapidly being acquired, we were enabled to impart to her a sufficient understanding of the proper address and salute for officers, telling her, of course, that no opportunity should be permitted to go by without saluting in "good English and proper form" all American officers, with the exception of Lieutenant House. Under no consideration, she was instructed, should she pay a "similar deference" to him. After our instructions had been assimilated by our pupil, according to our intentions, and while on detached duty one morning walking through the village street I saw our old Frenchwoman friend approach a forming company. She planted her heels together in true American style and with hand raise in salute, shouted: "Cap-e-tan: you beeg s— of a b—", and the captain's face went livid.

American doughboys, too, wanted to learn the language of the new country. There were many arguments as to the most efficacious manner of this accomplishment. Some concluded that John Paul Jones was right when he said: "There are two ways of learning French commonly recommended,—take a mistress and go to the comedy." As we did not have access to the comedy that method did not prove popular; however, the learning of French progressed with the passing nights.

Pay-day came and with it a holiday. Francs and centimes flowed freely on the spread blankets, the lucky one taking the winnings. And with the end of the day, despite a strengthened guard around the village, the boys began "to go over the hill" toward a little town "off limits". A guard of our own men was not difficult of negotiation for he who guarded also wished he might be free for the night.

I trudged down the road to the nearest oasis and, when the hour was late, started unsteadily back again to Colmiers La Haut. Our path led under a railroad, an underpass. There a stalwart half-drunk out of my company had located himself. He had been unlucky with 'Little Joe' that day and had no money. As the irregular stream of returning soldiers marched along a call would come from the under-pass darkness: "Halt! Who goes there?" . . . "An American soldier" . . . Limned against the skyline I saw a man with what appeared to be a rifle 'at the port" . I advanced, and the soldier with a stick for a rifle reached into my pocket and taking therefrom a flask, drank guzzlingly, and said: "Now I recognize you . . . Pass!" I did so, but lay on the top of a nearby stone fence and watched the 'guard' until he saturated himself with his liquid gleanings and ineffectually attempted to waddle his failing legs homeward.

Then came gas-mask practice, hand-grenade throwing, and bayonet practice. I was especially schooled in bayonet work, becoming rather proficient in the use of the blade within a short period. One of my non-commissioned-officer-instructors, who had previously evidenced a dislike for me, criticized my bayonet instruction to the commander, adding that the men were not receiving from me that efficient instruction necessary for proper offensive use of the bayonet. I told the officer I would be pleased to have the benefit of any superior knowledge the sergeant might have, and he was thereupon delegated to give me that instruction. During the first few moments before my antagonist (for such he had in fact become, I could see) I made a few lethargical passes with the bayonet, feigning, and barely warding off a thrust at me; and when the sergeant smiled a sardonic smile of satisfaction, I flashed at him and stuck him deep . . . And he went to the hospital . . . I always thought the captain understood.

Sergeant "Red" Hurmans was company mechanic upon whom fell the responsibility of loading our grenades,—not by any means the safest of assignments, for a mistake in handling the detonator, or dropping the grenade after loading, would almost invariably result in its discharge, usually with fatal results. With him worked John O. Lilley, a princely freckled faced chap. These men became experts at grenade loading, but after handling them constantly for weeks they naturally became a little careless.

As a part of our practice we were taught the use of the different kinds of grenades as well as the manner of handling them in battle. Two kinds of grenades were used by members of the A. E. F.,—the serrated grenades and the "eggs", as we named them. A serrated grenade was no different in construction from the 'egg" other than the covering of the serrated grenade was of cast-iron with grooves moulded into the covering while the "egg" covering was of tin. The serrations were for the purpose of causing the grenade to break into small fragments when the grenade exploded. These fragments, larger than the ordinary rifle bullet, would fly in all directions and inflict great damage on all in their path. They were intended only for use where they might be thrown a sufficient distance to carry out of range of the thrower. The "egg", however, when it exploded threw no pellets but exploded with a terrific concussion, a force sufficient to knock down any one near it. It is obvious therefore, that it was intended only for defensive (or close-range) work. For ex- ample, an "egg" might be thrown at an advancing German at close range, the German getting the impact of the explosion with no bad results to others nearby.

In learning to throw grenades, of course, the officers permitted us to use only the defensive grenade. One day I was called to the grenade-pit to learn to throw these bombs. John O. Lilley was loading for the men and Captain Whitaker was standing nearby supervising the work. As I came up for 'ammunition* there was none other than serrated-bombs loaded, and the captain criticized Lilley soundly for his failure to keep a supply ready for the bomb-throwers. With a leer, Lilley picked up a serrated grenade and shoved it at me. As I took it from his hand I saw it was not a practice-grenade and, of course, realized it might inflict serious injuries upon myself and other men occupying the area. Instinctively I hesitated before stepping into the pit from which I was supposed to throw the bomb and then take refuge behind the foot-high dirt parapet to ward off the effects of the explosion. Whitaker stepped to my side and said: "What the hell is the matter with you now? 'Fraid of it?" . . . Turning so he could see I had a 'mop-up' (another doughboy's name for a serrated grenade), I raised my arm to throw it when I felt the detonator-lever fly loose. Like a flash I knew it would be only a second until it exploded and I tried to throw it. When I did so, it slipped from my hand and fell just over the foot-high embankment. Whitaker exclaimed: "God help us!" and fell sticking the duck-bill of his 'tin-lid' in the dirt up to his face . . . and I was lying still flatter beside him! We waited a moment for the explosion, but it did not come . . . and he whispered to me: "What's the matter"? . . . Scarcely audible, I replied: "It's a dud." (A grenade which does not fire when thrown). He cocked his ear up, then peered cautiously over the parapet, and rising to his feet stalked away, calling back to me: "It's a hell of a place for a dud." We were not annoyed by his presence any more that morning at the grenade-pit.

Friday came again and with it, like the return of the bad penny, came our last practice hike. The day was warm, and we strained along with full packs keeping battalion formation. It so happened that L Company was in the rear and we kept that position until long past noon when we climbed the side of a mountain where we were permitted to sit and eat a sandwich which had been prepared for our noon lunch while the colonel harangued the battalion. He was able, however, to find little fault other than that he had noted the failure of the advance-guard to send out scouting parties before going into a village and the fact, as he expressed it "Your progress was inexcusably slow."

We reversed the order of march on the return trip. I caught advance-guard duty,—just what I had hoped for. I advanced the 'point' rapidly until the on-coming soldiers had sunk behind a hill. Then I called the men together, refreshed their memories as to the unkind criticism of the colonel, and asked if they wanted to 'walk the colonel to death' . . . Naturally they did. Thereafter it was "forward march, all speed". I sent a boy named Willie J. Jones ahead with instructions that he should run every time the battalion was out of sight. This he did, and we were soon far in the lead. Then Willie began to receive signals from the rear to "halt". I wig-wagged to him: "Double-time," and away he would go. The companies struggled to keep the pace. Finally the wig-wag came: "What signals are you getting?" Willie replied: "Double-time-halt."

With orders to stop at a cross-road, the 'point' sat upon the ground resting for a long time before the colonel, the major, and the captains with wobbly legs appeared upon the field. They had had the run of their military lives, but there was a gleam of good humor in the eye of the colonel when he asked the captain to find out from the corporal why it was he had tried to kill the whole damn battalion.

As we went into the little village, weary and hot, rain began to fall and we were wet to the skin. We had no extra clothing and were very uncomfortable, but a hot meal was waiting, we knew, and the men sang:

Ka . . . ka . . . Katie;
Ka . . . ka . . . Katie;
You're the only gu—gu girl
That I adore.
There's no moonshine
Over the mountain,
But I'll be waiting

I'll be waiting,
At the ki-ki-kitchen door."

Night came on and we were now wet and cold. The wineshops were swarmed, and when 'to quarters' was sounded by the bugler the tune had been improvised to sound like this:

Co-co-cognac!
Co-co-cognac.!
You're the only drink
That I adore:
Now the moonshines
Over the guard-house,
And I'll be mopping,
I'll be mopping
Up the ki-ki-kitchen floor."

"Lights-out" had long been sounded that night but to the wakeful the guards' steady tramp upon the cobbles reverberated through the streets to be broken now and then by some returning devotees of Madamoiselle, who sang:

"Madamoiselle from Saint Nazaire, parley vool
Madamoiselle from Saint Nazaire, parley vool
The Madamoiselle from Saint Nazaire
Never heard of underwear
Hinky-dinky, parley voo.l"

It was quite a treat for a platoon to be sent out alone with Sergeant Nash. He was a small man in stature and liked to do as little work as anyone of my army acquaintance. With him I was in full accord, but he was decidedly the more ingenious in avoiding contact with anything savoring of arduous effort. Leaving the company street one day with my platoon he stepped us out lively until we disappeared over a hill where he brought us to a halt for an "evening of rest". Concealed as we were in a scrubby woodland on the side of a ravine, he had picked a fine place for us to 'soldier'. Nash posted a look-out for the captain and we lay down and slept. When the day was waning, one of the out-posts suddenly signaled, "Look-out . . . Big Brass Hat!" and Nash, springing to his feet, called out: "That problem worked out all right, men . . . Now we will take a little rest. If the captain could have seen you this afternoon he would have been proud of your magnificent work" . . . and just then the captain came into view from behind a bush. The ruse must have worked, for there was no criticism, and we trudged homeward over one of the roughest pieces of cut-over land I ever saw in France, stopping at the spring to fill our canteens, feigning a fatigue which did not exist.

Having run the gamut of a full course of training, our time thereafter was taken up chiefly in intensive extended-formation practice. The importance of this training was: Massing of troops under conditions of modern warfare is dangerous because of the long range guns. When many troops were together the casualty list was the highest. By using the extended-formation,—separating troops into thin lines,—the intervening space between men lessened the loss from long range guns as well as machine-gun fire.

The practice of this formation was most exhausting to soldiers because they had to move more rapidly. It was therefore the most dreaded of formations. Near the close of our sojourn in Colmiers La Haut we were experiencing the most disagreeable of these maneuvers. Our practice had been so intense that we had gone stale. Our bodies ached and the men were in a bad frame of mind. The drill became 'sloppy', and the captain was especially severe upon his officers and it appeared to the men he never neglected an opportunity to force the iron of authority into our souls. One day I was running with King in an extended-order maneuver when we fell upon the ground together and King blurted out: "If that red muzzled would keep away from us we could do all right."

The captain appeared almost instantly at our elbows and commanded:

"Corporal, report at quarters tonight."

After chow-call that evening I met King going up the street, and he ventured the assertion that it was finally 'all off' with him. He disappeared into the P. C. but returned in a few minutes reporting the following conversation:

Captain: Corporal King, what was that you said today when we were at drill?

King: Sir: I don't recall. To what does the captain refer?

Captain: Your statement, corporal, when I told you to report here.

King: (Laughing) O, I recall now, sir, I told that red-snouted ?x;z?! in my squad to keep the muzzle of his gun off the ground or he'd blow us into the ditch."

Captain: Very well, corporal, I suppose I misunderstood you.

King's celerity of thought and action often brought him unscathed from very close places. This incident, so far as I ever heard, was never spoken of again.

About this time we lost our best liked officer. Needless to say it was Lieutenant John R. House. The friend of the soldier, the enemy of unfairness, a true soldier himself, never fearing to do the thing he required of his men, always eager to take the part of a common soldier against an officer, he had found a place in the hearts of the men of L Company which could not be filled by any man. Men went to their graves with his image engraved upon their hearts,—the true man, the real American officer!

Early one morning an orderly told me to report to the lieutenant's quarters. I found him lying in bed with all evidences of a physical sufferer. When I appeared at his door he asked me to come in, inviting me to a seat. He told me he had wanted to have a talk with me. Naturally I concluded he wanted to inquire into the persistent reports being circulated amongst the men which had led up to the near mutiny in the company, but I was put at my ease by his assertion that he 'just wanted to talk a little'. He stated he had been ill of late and had not had the opportunity to talk with some of the men; that he had asked me to come to see him to find out something about my education, etc. He then with great earnestness stated he had a profound interest in the welfare of the men, and wanted his noncommissioned officers to take care of the men under their charge. He then said an opportunity, most likely, would be given me soon to go to an officers' training school. I told him I did not care to go; that I preferred to remain with the men with whom I had trained. He expressed his doubts about the wisdom of my attitude, but, he said he could see my viewpoint, although he admonished me to remember that the best service any man could perform for his country was the use of intelligence properly directed and not the expenditure of physical effort. I gained the impression he wished to say more to me than he could afford to say . . . A few days later I understood.[1]

Sergeant James O. Allen received orders to take the fourth platoon to the drill-field. Hardly had we arrived until we saw Lieutenant House coming slowly toward us. He began to talk to us in a soft modulated tone, saying he was appearing before the company for the last time. At this point tears filled his eyes and he was so overcome with emotion he could not proceed. He turned and walked to the crest of the hill where he stood with his back toward us for several minutes. Finally he returned and expressed his regrets at his inability to control his emotions. He added that his fate had been unfair to his desires; that he wanted with all his heart to go with us to the front. He felt, he continued, that his having been one of our trainers had placed an obligation upon him to suffer with us anything which might be our lot; that it had been his ambition to be with

1 (Ed.) In 1918, the A. E. F. opened the Army Candidate School (ACS) to train enlisted men as officers. Originally about a three-month course, those commissioned from the school were called, "Ninety-day Wonders." Unfortunately, the heavy losses of Infantry Platoon Leaders beginning in the summer of 1918 until the Armistice caused the entire process of pre-commissioning training and Infantry Officer Basic Course shortened to a total of six weeks. The average survival rate was three weeks of combat. The idea was resurrected in 1941 and continues to this day as the Officer Candidate School at Fort Benning, Georgia. Joseph Douglas Lawrence. Fighting Soldier: The AEF in 1918. Edited by Robert H. Ferrell. (Boulder: Colorado Associated University Press, 1985.

us to the end. Then he admonished us to support his successor as we had supported him. "This", he continued, "I believe you will do. But there are two men in this company who, in my judgment, have failed to take their service in the proper spirit, and now is the proper time for us to have an understanding of ourselves. These two men have a wrong conception of your commanding officers. I will not, however, subject you to the criticism coming from public exposure of your names. Your own consciences name you. For your failure to perform with this company in the proper spirit I hold no ill-will toward you. I forgive you your personal unkindness, *and I wish for both of you a safe return to America!* I commend you upon your discharge from the army to good citizenship, hoping the experiences you will have in the army will show you the wisdom of cheerful co-operation! Some of you men will soon die. This is not a prophesy. *You are soon to go to the front, and nothing could be stranger than that all of you should come back alive*"...

Then calling us to 'attention' and with tear-stained cheeks he raised his hand in a last salute to the men he was leaving. Turning quickly, and with shoulders thrown back in true soldier style, he strode rapidly away... to the top of the hill... then his shoulders drooped, and his head was hanging low as he passed from my view... In this manner did Lieutenant John R. House pass from our army- life.[2]

The rumor... commonly called the "morning's poison gas" so named from the locale of its origin... had many versions of why the lieutenant had left us, but my conversation with him made me think he was physically unable to continue with the company. None, however, felt he had left the company through his own choice.

I had an experience about this time which all but resulted in this book having another author. I was sent out with a squad of men to do long-range rifle shooting. The idea to be carried out was: Firing over the crest of a hill from a position where the men could not see the target and striking the object in the valley below and beyond. This was accomplished, of course, by taking advantage of the trajectory of the bullet's flight. The

2 Lt. J. R. House now resides in Lincoln, Nebraska.

officer directing the firing would assume a position in front of the men and at a location where he only could see the object, and observing where bullets would strike from a trial shot, he would order sights set and destroy the target on the second or third volley.

I had in my squad that day a man who had been released from the Texas penitentiary (I afterwards was told,) on the promise that he would join the army. It was said he had been convicted of murder. Some days before our range-shooting I had been required to supervise the work of this man, and because of his obstinacy, was forced to deal with him at times rather severely. Therefore, he did not hold me in his highest esteem, but as it was all a part of soldiering to me, I did not have an idea he would attempt to murder me in revenge.

Arriving at our position on the side of the hill before the target in the valley beyond, I spread my men on the ground below the rim and took position above and in front of them directly in their line of fire. Here I lay flat on the ground and peered over the hill. The formality of this practice (the manner in which it would have been done in actual warfare) was carried out in detail for the reason that the hills were "all littered up" with spying officers who constantly held their field-glasses upon us, and few results went unobserved by them. If practices were not done in accordance with the methods "made and provided" then the non-com in charge could expect an unpleasant accounting that night.

I found my target, guessed the distance, called the 'guessed' range back to the men, and under ordinary conditions would have given the 'fire-order' without looking backwards, letting the volley pass over me. Why I did not do so, of course, I cannot tell, but I turned over and looked back, and this man, contemplating the order to 'fire', had leveled his rifle straight on my head . . . I could look right straight down the barrel of the gun! . . . Then I simply did not give that order to fire! . . . but crawled back to him with bayonet drawn and villified him to the extreme limit of my vocabulary,—and by this time, too, I had acquired some considerable of the indelicacies of the art! He simply lay there whimpering. The men had seen his intentions . . . Then I crawled back to my position . . . and summoning all the courage at my command . . . lying belly-flat on the ground, and holding my breath, not knowing what might be next, I called "fire." . . . And no bullet struck me! Guessing the range again, based upon my observations from the trial-shot, the second volley tore up the target . . . and I marched the men away.

I went back to my position because I was *afraid to be afraid* of this man. I felt that had he once 'bluffed' me he would have killed me later, and it was better to have the issue decided then and there. I made no mention of the incident but the men talked about it around the company and it must have come to the ears of the captain, for he asked me one day if I wanted this man transferred from my squad. I told him I did not. I kept him until we went to the front, but before the days of real trials came, he had become my friend. On the way "up" while he was sick on the road, I helped him along with his pack, and gave him a supporting arm. He responded with all the obedience his intelligence would permit.

King also had some 'incorrigibles' under his command who placed no value either on their, or other people's, lives, but he commanded their respect by protesting when the captain directed that they be transferred from the company, and begged that he should be permitted to keep them. He dubbed them 'The Sing-Bad Quartet', (probably their only real accomplishment,) and urged them into every 'Frog-fight' available, thus keeping up their 'fighting spirit' . . . There were some hard men, indeed, in that company . . . not many, but just enough to give the company a flavor.

Listed as one of the 'incorrigibles' was a little Italian boy. Naturally we called him 'Italy'. He came to us at Winneldown, having come over-seas with another organization. He had gone A. W. O. L. from his company, had been captured and assigned to L Company. He was the bane of the captain's existence. He never learned anything, but if I am mistaken in this, then, he failed to put his learning into practice. One day the commander told me to "take that Italian over the hill away from the company and see if you can poke something into his skull." Orders were orders . . . so off went Italy and Ireland. The short talk with him convinced me he did not understand the English language sufficiently to grasp the import of the commands unless spoken to very slowly. I got far enough away from the company with him . . . and of course, away from the commander, which was my primary purpose . . . so that what I was doing could not readily be observed.

I asked 'Italy' all about his people, where he had joined the army, where he lived, what he had been doing as a civilian, and in this manner worked up to an understanding with him of the importance of armies, regulations, and practice. He was able to understand the manner in which I conversed with him and soon he was asking me: "How do you do this?" . . . and this . . . and this? . . . I would show him and tell him the order in English. He would do it over and over for me, always repeating the commands aloud. Finally I sat upon the ground, and he became a "full parade passing in review" by himself . . . and I bragged on the results.

I was having a good time. So was 'Italy', but then the officer had to come sticking his nose over the hill. There he stood for some time looking at us. He came down the hill and asked me: "How in the hell did you do this?" I said: "Sir: 'Italy' does not understand English well enough to know what it is all about," but without taking the tip from me, in a deep-throated cynical manner he demanded: "Why in the hell can't you do this for me?" 'Italy' spread his hands and began talking . . . And O, how that boy could rattle Italian! . . . The commander demanded: "Now whatcha sayin'?" . . . and 'Italy' answered: "O, eet maka no diff. You no understande me eether."

General Allen's orders came for a battalion inspection and a division maneuver. I missed the inspection, having gone on guard, but heard from the boys when they returned that the general had 'poured it on' the captain for his failure to have some of the men properly equipped, especially with shoes. The demeanor of the boys at their billets that night reminded me of the funeral of the cat attended by the mice so 'distressed' were they over the chagrin of the commander.

On the division-maneuver I had the time of my life. I was fortunate enough to be with the rear-guard, a position which placed me far back of the advancing companies, and daylight revealed to me one of the most beautiful of spectacles. Spread out over all creation, as far as my eyes could see, were men, men, men . . . long wavering lines of men, moving majestically forward like the crests of the waves on the sea . . . olive drab against a background of green . . . and horses, glinting sunlight from their slick coats, tugging slowly . . . and wagons . . . big guns. The dust was rising lazily to obscure the march. I lay on the hillside and feasted my eyes on this man-made composite whole wondering just what part in this tragic drama of life and death I might soon be playing!

Editor's Notes

On page 102, Emmett includes the verse, "Automatics Right!," when recording a Doughboy marching song written and performed on the march by a fellow Soldier while returning from a field training session in France. This hint that Emmett adds to his tale is the first of many that his company was organized for combat to fight in the trenches using methods adopted from the French. Sung to the tune of "We've been working on the railroad," it notes that the captain was shouting "Automatics, Right" as a command and informs that he was directing the automatic rifle teams to advance in single file column on the right side of the half platoons in each platoon. These weapons and formations were new to Soldiers who were herded to ranges in Texas and into French villages for billeting and training as barely organized mobs. Note that Emmett and his comrades trained with the same company at Camp Travis, crossed the Atlantic on a troop ship, and trained for a period in France before being organized further into platoons. He states that after arriving in their last training area behind the lines in France, he was assigned to the 4th Platoon, 7th Squad, as a newly-assigned corporal and squad leader. Squads in the United States Army Infantry had eight Soldiers each. Multiplying eight by seven gives one fifty-six. Add a Platoon leader, a Platoon Sergeant, and two Half Platoon Sergeants, and one arrives at a total of sixty men, close to a full-sized fifty-nine man rifle platoon mandated by the new regulations. Emmett would have only a few short weeks to settle into his new responsibility, however. In fact, his regiment, the 359th Infantry, would billet in small villages and train in the fields around Recey-sur-Ource for five weeks, absorbing more new recruits who were not yet taught to salute and shoot, before moving directly into combat.[3]

Emmett was again promoted in the field from corporal to the rank of acting sergeant and placed in charge of a half platoon while marching forward into his first battle. Specifically, the Sergeant of the 2nd Half Platoon within his rifle platoon. In addition, he was asked to recommend his replacement as squad leader from among the privates of his squad. This

3 (Ed.) Infantry Drill Regulations, United States Army, 1911. Text Corrected to Include Changes 23, Sept. 10, 1918, and Appendix D, U.S. Rifle Model 1917 (Enfield), 19; Wythe, 14-15.

shows that the situation was rather fluid and Emmett was recognized as a leader when combat was eminent. He would not be alone in taking up a new task. The entire United States Army in France, and the one attached Marine brigade, was transitioning to new formations, weapons, and ways of warfare.

Shortly after the declaration of war, decisions were made by the United States War Department to adopt French weapons and formations for combat on the Western Front. While still organizing divisions and training leaders in the United States prior to going overseas to fight, the important manual, *Instructions on the Offensive Conduct of Small Units, Translated from French Edition of 1916, Edited at the Army War College, Washington, D. C., May 1917,* made its way to leader training camps and became the early basis for training what was to become the A.E.F. Included were versions of the new French formations for advancing through artillery barrages and machinegun and rifle fire across No Man's Land.

After the implementation of the trench system by the armies of both the Allies and Central Powers, all armies developed specialty weapons and formations to deal with the challenges of trench warfare, particularly with advancing though the fire of artillery and machineguns. The French, like the Belgians, were particularly motivated to figure out ways to force the Germans to withdraw from their occupied homeland. That would require methods to either cause a collapse of empires on the home front or to develop methods to push entrenched armies back across international borders. In three years of trench warfare, it seemed that there were no viable strategic or operational methods of moving large portions of the front line in either direction. A week to ten days of intense artillery bombardment, followed by a concentrated surge of hundreds of thousands of assault troops, failed to significantly move the front line at Verdun, the Somme, and everywhere it was tried. In essence, small groups of Infantry had to arrive intact inside the enemy tranches with enough weapons and leadership to hold against counterattacks and enable a continued advance. While general staffs struggled with solutions in the rear, the daily pounding by artillery of troops concentrated in trenches proved the greatest cause of casualties in the Great War and in every conventional combat since.

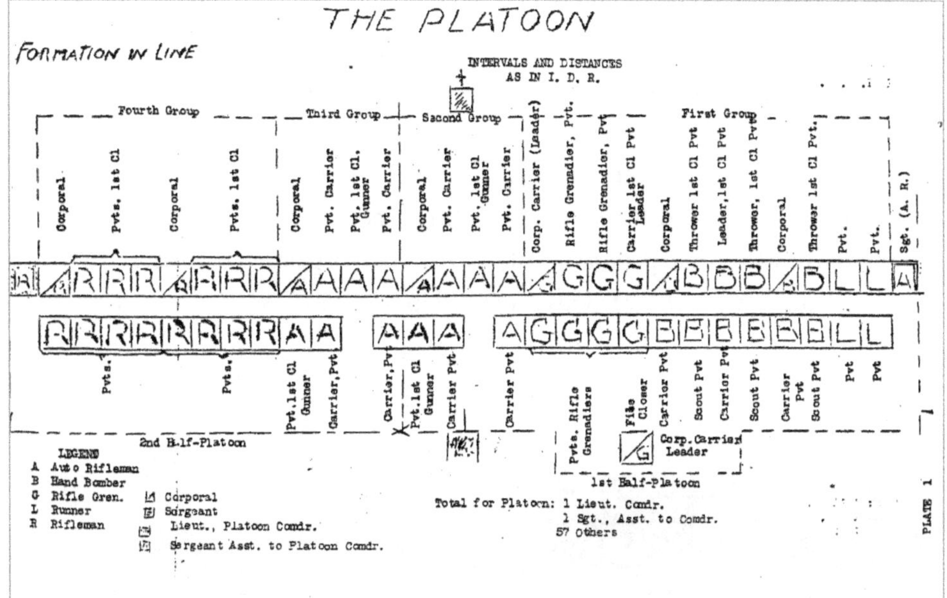

Original hand-drawn and typed copy of the early 1918 Rifle Platoon organization. This version is dated April 1918 and was the formation taught to units until after the St. Mihiel Offensive when it was simplified by many units. In this version, the two half platoons represented a support-by-fire element on the right and a maneuver element on the left. Refer to the text. Copy in Stieghan Collection.

The French settled on a means of empowering the Infantry platoon of around forty men to lead the advance through a reorganization of their formation and specialty man-portable weapons. In addition to the traditional rifle squads of around eight men led by a corporal, small teams and groups were organized to wield hand grenades (often called hand bombs), rifle grenades launched to 200 meters from cups attached to the muzzle end of rifle barrels, and the new light automatic rifle, the Model 1915 C. S. R. G. Chauchat. These light weapons were designed to allow Infantry Soldiers to carry explosives and automatic fire across No Man's Land and into the enemy trenches. They were calculated to be light and allow for the destruction of troops in trenches, bunkers, and to hold gains. Assaulting Allied troops with a lack of ammunition and losses of critical leaders would then have to break up the inevitable German counterattack. Those surviving units were expected to perform a position-holding mission when it was the weakest. While it was always costly in terms of lives to get across the open space between the lines and through the barbed wire obstacles, it proved more difficult to maintain a decimated assault force short on ammunition, water, food, and particularly, leaders, inside the newly-captured trenches. The holding attack became the approved method of capturing key terrain as a preliminary to continuing an advance. Though agonizingly slow, and costly, and moving a few kilometers each time every few months, it was the only method the generals could envision for regaining maneuver on a battlefield dominated by troops protected in trenches.[4]

Training of the post-graduate type in France, preparatory to entering the "big party" that is going on at the front lines. The picture shows an automatic rifle squad with a scout in their correct positions at an American military camp near a provincial town in Northern France.

U. S. Army Signal Corps photo of an Infantry platoon rehearsing open order formations behind the lines in Northern Franch. Each group of two is an autorifle team with a Autorifleman and his 1st Assistant marching alongside wielding a Chauchat Automatic Rifle.

On January 14, 1918, the United States Army officially adapted a new regulation that reorganized all Infantry units. From 1778 to 1918, the rifle company was the smallest permanent organization of American Infantry troops with 100 Soldiers. The captain commanding was armed with a pistol and all the others carried rifles and bayonets. With the stroke of a typewriter, the Army adopted a new French-type company that grew to 256 men and was now subdivided into four rifle platoons and a headquarters platoon.[5]

4 (Ed.) Army War College. Manual for Commanders of Infantry Platoons. Translated from the French (Edition of 1917) at the Army War College. Washington, D. C.: Government Printing Office, July 6, 1917.

5 (Ed.) "After a thorough examination of allied organizations, it was decided our combat division would consist of four regiments of infantry of 3,000 men, with three battalions to a regiment, and four companies of 250 men each to the battalion…". Pershing, General John J. 1919. "General Pershing's Official Story; Battles Fought by the American Armies in France from their Organization until the Fall of Sedan." Infantry Journal XV, no. 9 (March): 692.

Each of these rifle platoons contained fifty-nine Soldiers and were led by a Platoon Leader (lieutenant) and a Platoon Sergeant. The platoon became the smallest tactical unit and was organized into two half-platoons, each led by a sergeant. Note the illustration provided from an original hand-drawn training document found in an Infantry sergeant's trunk from the 79th Division. The 1st Half Platoon on the right was composed of teams and groups of Soldiers armed with hand bombs (grenades), rifle grenades, and two automatic rifles. The 2nd Half Platoon on the left included the other autorifle squad of two teams, and two eight-man rifle squads. The two halves of the platoon were designed to fight in tandem as a support-by-fire element and a maneuver element. The platoon leader could use the 1st Half Platoon of his unit to keep an enemy machine gun occupied in front with rifle grenades, hand grenades, and a few autorifles, while the 2nd Half Platoon would maneuver under cover to a flank to destroy enemy machineguns with grenades and bayonets. It was a machinegun-killing machine.[6]

Original posed photo of an Autorifleman with his Chauchat at the edge of a frontline trench. The detached nearby magazine still has a few loaded rounds and a few fired cartridge cases are on the ground. He has drawn and aimed his cocked automatic pistol, a U.S. Model 1911 Pistol of the type Emmett was issued in France. Note the stakes placed in the trench to impale unwary attackers as Emmett encountered at St. Mihiel. Stieghan Collection.

6 (Ed.) Table 7. Rifle Company, Infantry Regiment, Maximum Strength. United States Army in the World War, 1917-1919. Organization of the American Expeditionary Forces. Volume 1. (Washington, D. C.: Center of Military History, United States Army, 1988), 347. Instructions for the Offensive Combat of Small Units. A copy of the April 1918 training circular, signed by General James G. Harbord, Chief of Staff of the A. E. F., and approved by General John J. Pershing, Commander of the A. E. F., was provided to the author by a private collector in 1995. The document is hand typed and the formation illustrations are hand drawn. See the accompanying illustration.

All armies on both sides adopted similar combat group formations by 1918, though the Germans perfected it as entire battalions of "storm troopers" to infiltrate an enemy trenchline through stealth to penetrate by small groups deeply into Allied rear areas. The advanced guard of the five German offensives of the spring and summer of 1918 were led by storm troop battalions that penetrated on one occasion fifty miles in one week in the British sector. They were stopped by resurgent Allied armies joined by newly-arrived American troops who helped save Paris from capture by Soldiers of the United States 3rd Division digging in at a bend of the Marne River and Marines who crossed wheat fields to counterattack into Belleau Wood.

As a result of the first few months of combat experience, American leaders began to fine-tune their rifle platoon formations to make them more flexible and easier to lead with inexperienced leaders and replacement Soldiers. The earlier *Instructions* manual was updated by Pershing's staff in France and published in Paris in Spring 1918. *Instructions for the Offensive Combat of Small Units*, appeared in April and was distributed down to the platoon level for training and use in the summer battles in France. A month later, it was also printed in Washington, D. C. by the War Plans Division for use of leaders to train divisions that had not yet left the United States for the Western Front. Included were illustrations of the new rifle platoon of two half platoons and the various means of advance across the battlefield and deployment under fire.[7]

The General Headquarters, A. E. F., published a small booklet, *Combat Instructions,* in Paris in August 1918 filled with fine-tuning directions for leaders of small unit leaders as a result of battlefield observations and

7 (Ed.) War Plans Division. Instructions for the Offensive Combat of Small Units, Prepared at the General Headquarters, American Expeditionary Forces, France, from an official French Document of January 2, 1918. War Department, Document No. 802, Office of the Adjutant General, May, 1918 ;See also, Headquarters American Expeditionary Forces, France, Instructions for the Offensive Combat of Small Units, Prepared at the General Headquarters, American Expeditionary Forces, France, from an official French Document of January 2, 1918. A[jutant] G[eneral]. Printing Department [Paris], G. H. Q. A. E. F. [General Headquarters, American Expeditionary Forces], 1918. A copy of the March 1918 training circular, signed by General James G. Harbord, Chief of Staff of the A. E. F., and approved by General John J. Pershing, Commander of the A. E. F., was provided to the author by a private collector in 1995. The document is hand typed and the formation illustrations are hand drawn. See the accompanying illustration. The editor has a copy of a hand-typed and hand-drawn version of the essentials of this manual discovered in an Infantry sergeant's footlocker from the 79th Division. It appears the sergeant's unit could not get enough copies of this important training circular and hand copied enough for their important leaders to learn and train their Soldiers.

reports. It urged leaders from battalion level and below to simplify their formations and to get Soldiers to spread out more on the battlefield to prevent needless casualties from machineguns and artillery fire.[8]

Original posed photo of a German Stosstruppen, or "Storm Trooper," trained and armed to infiltrate across No Man's Land and entering Allied trenches armed with grenades and a carbine slung across his back. Note the barbed wire cutters hanging from his belt, the leather patches protecting his knees while crawling, and that he wears shoes and has tucked his trousers into his socks instead of wearing the standard tall boots of the German Infantryman. Six explosive heads are unscrewed and wired around a single "potato masher" hand grenade. Stieghan Collection.

8 (Ed.) Combat Instructions. A. E. F. No. 1348. GHQ, AEF [General Heqdquarters, American Expeditionary Forces]. Chaumont, France, 5 September, 1918; Combat Instructions, A. E. F. No. 1348, War Plans Division, October 1918. Also see: War Department Document No. 868, Office of the Adjutant General. War Department, Washington, October 5, 1918.

Small units across the A. E. F. began to experiment with alternate formations to make maneuver simpler and to train groups to continue to fight regardless of leader casualties. In 80th Division rifle companies, for example, both half platoons were organized as identical groups with the same squads, groups, and sections containing equal numbers of assigned weapons. The hand bomber, or hand grenade, teams were omitted and each Soldier was issued at least two grenades, instead. Another formation illustration appeared in an article on half platoons as used in combat published in the *Infantry Journal* by a combat Infantry battalion leader in April 1919:

> In the Soisons-Rheims offensive attack formations of platoons, companies and battalions were too dense and followed too rigidly the illustrations contained in "Instructions for the Offensive Combat of Small Units." Waves were too close together and individuals therein had too little interval, columns were lacking in elasticity and little attempt was made to maneuver.
>
> A close study of the best means to correct these faults led to greater emphasis being placed on the half-platoon as an elementary unit. Experiments conducted in rear areas developed the formations illustrated which were utilized in the last Argonne offensive and thoroughly justified their efficacy by greater maneuver power, better control, rapidity of deployment and conservation of life.[9]

A few weeks before combat ended with an armistice, the United States War Department arranged to have a new manual published in Paris, *Infantry Drill Regulations, (Provisional) American Expeditionary Forces, Part I, 1918*. The thin paperback version began printing on December 12, 1918, and copies were rushed to the front. By the time these pocket-sized booklets arrived in the hands of leaders in the AEF to train troops in the newest attack formations, the shooting ceased with an armistice and the Doughboys had reached their occupation stations along the Rhine River

9 (Ed.) Major Henry H. Burdick, 318th Infantry [Regiment, commanding 1st Battalion], A. E. F. 1919. "Development of the Half-Platoon as an Elementary Unit." *Infantry Journal* XV, no. 10 (April): 799-809.

inside Germany. Surviving original documents used to train the Soldiers and Marines of the AEF in the way they fought on the Western Front are rare and scattered in the papers of junior leaders. Emmett's memoirs show that his regiment, at least, was among those organized into half platoons and fought that way in the St. Mihiel Salient offensive.[10]

An unidentified group of Doughboys taking a rest break while on a road march. They have full packs worn different ways that include white paper-covered packages of hard Army Bread. Note the non-commissioned officer in the right rear with hands on his hips and a wound stripe on his sleeve. Stieghan Collection.

10 (Ed.) *Infantry Drill Regulations (Provisional) 1918, United States Army.* General Headquarters, American Expeditionary Forces, Paris, Imprimerie E. DeFosses, 1918.

Chapter 7

Swirling Toward the Maelstrom

About the tenth of August we received orders to police up the village very carefully. I had recollections of Camp Travis; the scene at Camp Mills returned to me; and I knew our training period had come to a close. An early morning bugle brought us from our uncomfortable beds. Scurrying feet warned us this was to be a day of events. We rushed our breakfast and reported to formation with every vestige of our equipment. The long lines looked shabby from over-coats tied at every angle to the packs. We were then commanded to stack over-coats for truck transit, but when a big disheveled pile of coats rose up in the street, a countermanding order came . . . not a thing to be vexed about in the army! . . . and we were required to find our over-coats in the big pile and wear them . . . Thus equipped we stood in line . . . impatient, stamping . . . murmuring . . . and everybody was guessing "Where now?"

A methodical count of the company was made. "All present and accounted for", reported Sergeant Meeks, and from far down the street echoed Whitaker's last command in Colmiers La Haut:

"L Company . . . Forward . . . March!"

Raking, rasping rumble of marching feet, crunching stones . . . left . . . right . . . left . . . right . . . tramp . . . tramp . . . and we were going away . . . away from the little French village we had learned to call "home" . . . from the little French girl with the long curly flaxen hair . . . from the old church and its bell that peeled so plaintively each day . . . from the abject old folks who moved with bowed shoulders, who with supporting sticks in hand tottered out to green pastures each morning with lowing cattle . . . from the bleak old stone walls . . . old, yes, so old! Yes, we were leaving . . . and we knew 'any home' was better than the march. But the spirit of adventure welled up to blot out the past . . . and we were headed toward the future.

But my mind lapsed back to reality as we made a column-right and headed northward toward the open country. The sexton in the old stone church pulled the bell-cord announcing our departure and the bell seemed to speak in unison with our foot-steps . . . "gone . . . gone . . . gone!" Far down the little lane which led alongside a rock-fence, converging khaki uniforms began to form a flowing stream of olive-drab . . . And then I was amused! On top of the rock-fence lining the road-side sat the entire feminine population of the city!!!

Cheers . . . "Goodby" . . . "Au revoir" . . . resounded from every throat. The company broke into wave after wave of reverberating cheers . . . calls . . . "Goodby, Madamoiselle" . . . "See you toot-sweet" . . . "Madam!" . . . and our feet crunched under the heavy loads as we picked up the cadence and swung down . . . down . . . the lane . . . And over the fence came a little head . . . flaxen curls . . . a small piping voice . . and someone lifted her and she stood upon the fence . . . waving . . . the little company sweetheart! She was waving frantically . . . throwing kisses to all who passed . . . crying, sobbing as if her heart had broken . . . "Goodby, Keeng (King): John O. Leele . . . O'leelee." . . . and a rising mist of dust scuffed up by hobnailed feet settled a smudge of brown on every face . . . on every tear-stained face in Old L Company! . . . But . . . I wiped away blinding tears from my eyes and stared! An apparition had appeared! . . . There . . . standing again at 'attention' . . . hand at salute, was our old Frenchwoman . . . "Cap-ee-tan . . . you beeg ?x !z! . . ." Drawn faces marked with emotion now glowed with laughter, and the 'Sing-Bad Quartette' pitched a high key:

There's a long, long, trail a'winding.
Until my dreams all come true
The nightingales are singing . . .

thump! . . . thump! . . . thump! . . . and Colmiers La Haut is no more a reality.

An ominous silence gripped the marching men. The grim realization marked itself on their faces. The drollery of practice was over and action now was to be ours. Wool socks, hot feet, heavy over-coats, and a shining summer sun forced our minds to present pains. Troops continued to converge upon every cross-road, and we walked four kilometers with only four minutes of rest . . . to take our place in the long line of march with the men of the 90th.

The column hesitated at a cross-road. Bronzed men, no longer bearing any resemblance to the rookies of Camp Travis, walked up and down the newly arriving company seeking for a friend. And there was old Joe Templeton, now, strong and rosy from the weeks of intensive training, no longer the fat boy of months ago . . . and I put my arm around him . . . and 'did he have a cigarette'? . . . but the column was moving again . . .

I had to hurry for someone at last had used some judgment. The company had orders to discard over-coats which were now to have a berth on a passing truck. It was a great relief to lose the coat, but the order had not come before streaks of pain had flitted up and down the backs of over-burdened soldiers. Down the road we went again to traverse another ten kilometers. An order passed along the line to 'eat dinner'. Food had not been provided for our company. Our kitchen had left ahead of us during the previous night in order to be in advance of the moving column and we were without food. We learned later that the team drawing the field-kitchen had run away destroying the kitchen and killing one of the horses. But King took the order " to eat" for just what it said, and he cut open his 'iron rations'.

The food carried by the soldier became known as the 'iron ration' because it was sealed in tin. This ration is never to be eaten except on specific order or after the greatest privation. A passing officer saw King enjoying himself and he was reported to headquarters. Later on,—just before we got to the front,—King was court-martialed for his 'offense' and I was called as a witness. I stated I had heard the order passed to 'eat' and that Lieutenant Nysewander had said, so it could be heard both by me and

King, that "It looks like iron-rations today". King was acquitted.[1]

During the noon hour of rest I went to a nearby stream and soaked my burning feet in its purling waters. I felt better during the long weary trudge which brought us, at the close of the day, to a small village through which we skirted, traveling through a street canopied by the graceful branches of tall green trees. The sun was still shining and the deep shade cast a spirit of restfulness over us. There was no boisterousness here as we walked along, but weary men raised their heavy iron-hats, swinging them on their arms, while they wiped away the grime from dust-covered brows.

A turn to the right headed us from under the tranquil shades across a stubble-field now yellowing in the summer sun, and we pitched pup-tents on the brow of an eminence over-looking a pastoral valley from which filtered up to us the tinkling of bells of slow moving cows being driven quietly homeward by peasant women. Our minds flitted back to similar scenes in the little home we had left early in the day, and we thought that there, too, our bovine friends were returning from open fields to lie down at the door steps of their owners . . . but . . . and we chuckled . . . we would not be there in the morning to be handmaids to their despoliation!

Rising below us was a great stone mound, probably the ruins of an age-old castle which had succumbed to the destruction of previous wars. And just as the sun sank to rest behind the little village, our band, now mounted below us on the old ruins, struck up "America". Instantly twenty-nine thousand men arose to their feet in salute. The last note trilled across the valley and echoed back to us only to die away . . . Silently the men from Texas and Oklahoma sat down. *They were thinking of home!*

This impressive scene caused our old gang, Tip, Bill, Molly, 'Elsie' and Joe to congregate before my "kennel-door." And we sat upon the ground until the moon had risen high into the heavens to look down upon a field of miniature tents whilst the tired thousands recuperated strength for the coming days . . . There was talk of home . . . of tomorrow . . .'and then to

1 (Ed.) Each Soldier carried a single tin with emergency, high caloric food only to be opened and eaten in an emergency. Also, a number of cans of "bully beef" (corned beef), "monkey meat" (corned beef hash with diced potatoes or carrots), stewed tomatoes, "goldfish" (salmon), "beans with pork" (pork and beans), and sometimes other canned vegetables were to be opened and eaten at the front only when hot, fresh food from the company kitchen wagon was not available.

bed on the cold hard ground . . . and . . . Tip had held his last meeting with the old gang! Only once after that did I ever see him.[2]

Our progress the following day was slow. Men, worn by the grueling strain of over-marching the first day, lay down by the roadside and refused to move. --- ---became sick and staggered to a fall. Officers cursed him. I took his pack and . . . well, I thought he was sorry he had intended killing me back on the range when we fired at long-distance targets. Others staggered along flinching with the anguish of pain from blistered feet. Still others tried to present a good front by an interspersed song:

"Keep the home fires burning,
While my feet are churning
Up and down the roads of France
We'll soon be there . . . ,"

and some leather-lunged fellow would call out: "Cut out the sob-stuff", changing the tune to one which would drown the first:—

"We've been drilling on the hillside,
All the livelong day,
We've been sipping at her fireside,
A sleeping on her hay
Lordy, Lordy what an army,
We'll make old Heine pay . . ."

but a town loomed up in the valley far below us . . . and the marching steadied down to a grim determined grind. A halt was called in a field a few hundred yards from scattered stone-buildings. We were at Vesoul with nothing more to do that day.

A hasty reconnoiter quickly brought old chums together . . . There was Bill, of course! And Red Rankin, John O. Lilly, and, of course, 'Judge' Goen

---

2 (Ed.) Since the Civil War, the small two-man tents buttoned together to make one were referred to as shelter halves. The Soldiers called them "dog tents" with a "kennel-door" because of their small size. Only one end was buttoned shut until new ones were issued with both ends closed during World War II.

approved our action . . . I made a run, passing an M. P. who threatened to shoot but didn't, and entered a French 'Off Limits'. I had little time to await arrival of the others. We ordered 'Three Star Hennesy' . . . but our officers came after us. . . . "Just orders, boys." . . . But we returned with bulging pockets and stopped behind a Frenchman's barn . . . Two officers came around the corner. We started to hide our bottle but one was a 'medico' . . . and he laughed at us . . . All of us then drank together and felt better.

Bill became convivial. I lay on the ground basking in a pleasant sun, but Bill would not have it so. He aroused me . . . and with the entire 'gang', I followed. Surrounding a 'Frog' who had been following with his two-wheel cart filled with wares for sale, (for which he charged a price that grew higher and higher with each kilometer,) Bill commanded: "Grab!" We did; then scurried! The 'Frog' went out of business,—we hoped, forever!

Then the 'good news' caught up with us that Elsie had let the horses run away with the kitchen, destroying it and killing one of our horses . . . The boys expressed no regrets over that, for they hoped that if they were attached to some other company for rations they might fare better . . . but they marked down the horse as "Casualty No. 1."

Night came along and Captain Whitaker was made 'loading officer'. Side Door Pullmans were waiting for us again and the movement began. What a congestion? Horses were moving in one direction and men moved in another. Long lines of men . . . cursing men . . . but the cars were filling. It was now too dark to see, so we moved by holding to the gun swung from the shoulder of the man next ahead,—a trick learned here and much used ever afterwards. In the congestion I heard a voice call out: "Catch him," and someone ran into me. I did not recognize him, but he slipped me a bottle of cognac and disappeared into the night: "Drink it. They are after me." Although I was heavy laden, I took on the extra burden thinking I might need it during the night to come.

Loading in the darkness under orders of one inexperienced in supervision resulted in the cars being filled irregularly. Some had a load to over-flowing while others failed to have their 'forty hommes'. It was my misfortune to get an over-crowded car, and sensing the probable situation to be endured that night I took a 'swig at his bottle' and backed up into a corner of the car and kicked off every man who trespassed upon my preemptions. Naturally a battle-royal was waged for a time, but being flanked on two sides by the sides of the car and on the inside by 'Three Star Hennessy', I held my possessions, finally getting enough space into which to crouch, where I spent a miserable night.

'Bull' Durham and John O'Bar went onto the roof of the car and rode there in the cold air. It was truly cold,—but they probably suffered less than I did packed below.

Morning found us at Toul. Our orders were to march to Pagny about four kilometers northwest. No breakfast, no water, and now I was feeling the effects of my dissipation of the night before. A high hill loomed up before us. We wound out of the valley to its side and stopped in the streets of Pagny. A wreath of green, flecked with purple, wound itself around the doorway of a Frenchwoman's home. We plucked, and the grapes were luscious. A jibbering old lady came to the door spitting vituperations as she shook her fists at "ee beeg dam Yanks". We gave her centimes and she smiled her invitation to continue our depredations.

We were assigned billets, and I was more comfortable than I had been at any time since leaving Camp Travis. For a time I occupied an upper section of a stone house in the lower part of which resided a French family. I had a wire cot. We improvised a table near the window which opened toward the mountain immediately before us. Bill Thompson, Singleton, Clarence Oliver, Kenneth Dunn, and I, now, were having a real good time as bunkmates.

A clear warm day with scarcely a speck in the sky . . . I sat at our point of vantage at the table. My eyes were directed heavenward, to the azure blue above the mountain top, and faintly there came to my ears the purr of a motor. I listened. It sounded like an air-plane. A raucous cry came up from the street. The French women were calling, "Boche" . . . "Boche." I peered into the blue above.[3] A tiny plane with the sun glinting from its

3 After the air-fight I continued my writing at the window, describing the incident as it is above. After I returned to the United States I saw that letter. The censor had clipped my description. They did not let the folks at home know what was really going on.

wings came into view. I watched it. It was a German plane emblazoned with the cross of black. Now he circled nearer and nearer and the motor throbs rang out his warning . . . He was dropping something. "Boom!" "Boom!" His bombs had failed to register a hit on the town, but he was paying his early compliments to the arrival of American troops. Anti-air-craft guns quickly picked him up firing salvo after salvo after him. Finally he was put in the 'pocket' with the breaking shells encircling him. He swerved, circled, dived, headed skyward again, trying . . . trying . . . trying,—O, so hard,—to get away! I craned my head out the window enjoying the fight as long as I could see him. And as he dived and escaped southward I could not but admire his unquestioned courage. The distant boom of the guns from Toul now punctured the air with their clang and I knew they were after him too.

When night came Bill Goodson and I went to the mountain-top to see the flares being sent up on the front now eighteen miles away. We sat on the very highest point several hours looking out into darkness. Every few minutes a flare—a Verey Light—would spurt into the sky and hang motionless for what appeared to be minutes before shooting into a myriad of stars. The low, distant, guttural grumble of guns could be distinctly heard. Invagination began to play . . . We wondered what it looked like up there! Did some one die every time a 'big dog' growled?

We were returning to our billets in the village below when our leisurely walk was interrupted by the murmur of an approaching Boche plane. Again the cry rang up to us from the villagers: "Boche . . . Boche . . . The Hun!" Then great streams of light burst from the mountainside. Searchlights were peering into the darkness in an effort to locate the raider. A brilliant shaft of light swept across the sky and . . . high in the air it caught him . . . a small bird with its head pointed heavenward, climbing,— climbing higher and higher,—trying to rise away from the light. Again there was a withering deluge of gunfire. Puffs of smoke . . . like burst of pop-corn, grey, then turning fleecy white . . . appeared all around the invader, but he dived . . . Darkness enveloped him, and we thought he had gone, but the deep intonations of exploding bombs came back to us. We knew he had registered on his target,—Toul. We arose from the ground where we had been lying flat on our backs viewing the entrancing pageant, and Bill remarked: "Well, the old bird has laid her eggs, anyway!"

Two months later, when again in this same town, a similar scene had a far different significance to me and Bill. We had had experiences on the front, had seen multiplied hundreds of similar situations, and a raid over Pagny meant more than a spectacle to us. And when a raider came over and there was the usual call of "Boche," American feet scurried with the French toward places of shelter . . . And Bill ran into me . . . but without tarrying, he shouted back: "Well, I've learned something since I was here last."

Several days of rest followed our arrival at Pagny. Joe Templeton and I put in our hours of leisure familiarizing ourselves with the places where we could purchase the best food, and it was the usual thing for us to miss 'chow-call' to sit at some Frenchwoman's table devouring euffs (eggs) with a bottle of vin rouge while the other fellows fell back into the chow-line for "seconds." Bill went along with us one morning to a new euff depot where we were met at the door by a comely young Frenchwoman. To her we explained our wants in our combined meagre French supplemented with many gesticulations. She stood with a smile on her face until we had exhausted our vocabularies, then she asked in perfect English: "What will you boys have? Eggs?"

The 'cooties' made their reappearance here. It seemed that everything the French touched became contaminated with vermin. I was walking down the street one evening when I noticed a peculiar greyish cast to the hair of a French child. Going closer to see the cause I saw it was vermin working in her hair. We had been reasonably free of cooties at Colmiers La Haut, but now we were forced to take every precaution to ward against them. The barber with his horse-clippers, therefore, became the most popular man in the company. As American citizens there were many styles of haircuts, but in France we had only one pattern—that taken from the old song:

"Good morning, Mr. Zip . . . Zip . . . Zip!
With your hair cut just as short as mine." . . .

But as we trod off in search of the barber, we'd sing:
"Good morning, Mr. Zip . . . Zip . . . Zip!
I'm certainly feeling mine.
Ashes to ashes, and dust to dust,
If Fatimas don't get you, then the cooties must."[4]

Joe saw me coming away from the barber with a pate pared and peeled. He would have turned away had we not brought physical pressure to bear upon him, and when he came 'out from under' there were about as many nicks cut in his scalp as we had cut hairs from his head . . . He was our friend, however, and we went away to a café arm-in-arm.

Pagny by this time was beginning to look and feel like home, but there could be no home for a soldier, so we began to talk that it was about time to move. Then one evening just as the mountain-shadow cast its gloom over our little village we were aroused to action by the shrill whistle of the first sergeant: "Fall out. Full packs."

Into the street rolled a long line of trucks. Pitch darkness came. I was abreast of a truck when some one called: "What company, soldier?" I answered: "L-359". The reply came back: "That you, Chris?" It was Jim Alexander who was lost in the shuffle back at the Camp Travis hospital . . . and now he had caught up with us . . . in the motor-transport service . . . away over here in France . . . But we could not take time to talk, for the infantry companies were marching . . . The moving trucks, hugging the center of the streets, were shoving us over to the right . . . The wind was blowing . . . and black clouds drifted angrily across the sky . . . and we were out again on the road trudging toward Toul. I was disappointed for I wanted to talk with Jim . . . but the wind cut my face and I forgot my grievance and tried to cover up with the lapels of my coat. They would not let us stop at Toul. The column hastened on northeastward. The line of march took us past many soldiers lying off to the right of the road in the mud. Horses were hitched to heavy wagons . . . The drivers were cursing. Many cannon went along with the men in the road. Tractors snorted with

4 (Ed.) Cooties were the ever-present lice that infested the wool clothing of the Soldiers of all armies until the introduction of steam boiler "cootie cookers" at the field divisional baths in World War I and DDT in World War II.

heavy guns trailing . . . and the rain fell, wetting us to the skin. We knew we were headed for the front for the Verey Lights constantly raised themselves high in the air directly in front of us . . . but many miles away . . . and they hung there an interminable time before showering their stars . . . Then the darkness was darker still.

Out of the din of moving men there came a spewing, hissing, sibilant sound . . . louder and louder . . . The air flared up with a streak of yellow light . . . Commands . . . high pitched calls . . . "Gas! . . . Gas!" I adjusted my gas-mask to 'alert' position and waited . . . sniffing the air. Again the call came . . . "Gas!" . . . and there was a commotion ahead . . . But it was too dark to see what had happened . . . I shoved the mask onto my face, clipping the nose-piece over my nose . . . and staggered on. O! how I wanted air . . . fresh air . . . just so I could get enough air to breathe! There was a rumble of a motor . . . I knew it was an airplane . . . and it was coming our way . . . "Fall out to the right . . . Under cover . . . Pass it on," . . . and we lay down in the mud there under the trees . . . Nearer and nearer came the motor's whir . . . All was quiet . . . except the motor throbbing up there. . . . Then the sky began to clear . . . The billowing clouds rolled back and the moon peeped through . . . And there was a burst of light . . . another . . . and another,—red, white, and blue rockets were being fired from the plane . . . We knew that signal. The United States of America was watching over us! We forgot the mud, threw off the gas-masks, got back in the road and went on.[5]

Now the night was clear. The storm had passed westward, but the wind blowing from the east across the front carried the muffled fire in deeply intoned grunts regularly to our ears. But the sounds were so far away we knew we could not get there tonight. Houses appeared dimly before us. We struck our heavy feet sharply against paving-stones. It was 2 a. m. by my wrist-watch which glowed up to me in the faint moonlight. We halted . . . "We billet here tonight" . . . What a relief!

We stood in line for an interminable wait. A Frenchman appeared,

---

5 (Ed.) The Small Box Respirator gas mask that the United States military adopted from the British had a wide shoulder strap with adjustments for two lengths. The long length was for a "carry position" across one shoulder to the opposite hip. When approaching the "gas zone" the gas mask bag was moved from the hip to the chest and the strap shortened to go over the back of the neck with a cord tied around the waist to keep the bag snug against the chest.

gesticulating: “Sorry, corporal, there are no more billets . . . except the guard-house.” . . . “The seventh squad, Sir,” I said, “has no mauvais honte.” (false modesty). He chuckled and we marched with him. The long wait had stiffened our unflexed muscles. I fell and a flagstone cut deep into my knee-cap. The blood oozed warm and comfortable down my right leg . . . But there was the guard-house . . . and sleep . . . and then . . . O! so soon, ‘chow-call’!

We scrambled out of the guard-house to peer out at a lowering sky. A forming chow-line greeted us. A passing buddy told us L Company had long since eaten. That was serious . . . I cast around for an expediency. Again my eyes found the nearby chow-line. An officer stood at the kitchen checking those who came to eat, weeding out those not assigned to the company, and I saw I was headed for trouble, but an obliging soldier told me the company was ---358. “What company, corporal?” . . . “Sir:---358” . . . “When assigned, corporal?” . . . “Just this morning, Sir.” . . . “Pass” . . . and I added sotto voce, “Just for chow.”

When we again joined our company we received quite a pleasant surprise when instructions were issued to billet us in a comfortable, clean, two-story building. Although I had to sleep on the floor there had been nothing like it since we left Camp Travis. Dumping my pack I struck out to find Joe Templeton. A bond of genuine friendship had grown up between us and we wanted to be with each other as much as possible. I found him coming “on the jump” to find me. He had good news to report. An old French couple had invited “the good American Yanks” to have dinner in their home. He had claimed the distinction of being “the best American Yank” and had recommended me also. We arrived at the hospitable French home with a beefsteak under one arm and a bottle of good wine under the other. Wrinkled old visages met us at the door with bows most ostentatious. There was a great clapping of hands when they learned we would contribute to the repast. The dear old man was a genial soul but our communication was seriously handicapped by the lack of a common tongue. He radiated his welcome, making us feel he had extended the invitation to the “bon homie Yanks” (good natured Americans) because the 'bete noir" (the black beasts,—the Huns) had taken his two sons, and he wanted to feel his sons were with him again. And he would "love the Yank with the beeg fat belly like my son" but there was no love left in his heart because of the Boche.

Our dinner was soon before us. The old wrinkled woman looked across the table with tears in her eyes. We could not make her understand our language, but her eyes spoke a language we understood. The dear old folks protested our arising until he had gone "to the cellar" although we had drunk one bottle of wine with our meal. I assisted his faltering steps into the darkness below where he searched amongst many bottles. Ascending, he brought one covered with dust. With great ceremony he pulled the cork . . . and indeed it was fine.

The afternoon was a clear one and orders came that we might go to the river for a swim. Thousands of naked men swam in the river while French women and children came in great numbers and sat at the water's edge on both sides of the stream. The river was wide but the water was shallow, however, boats loaded heavy constantly moved along with men rowing . . . "whack!" . . . "whack!" . . . "whack!" . . . as the oars were dipped into the water. We were seeing a part of France's system of transportation.

'First call' with the coming of another day brought with it some disappointments. We had orders to drill again. An effort was made, but scarcely had we reached the open-field until 'Jerry' could be heard whirring in the air. Lieutenant Nysewander said it was a Frenchman, but those who had been on the hilltop at Pagney caught the irregular humming of the motor and knew that 'Heine' was venturing far afield today!

He came direct toward the exposed troops. Then he seemed to develop motor-trouble and was losing both altitude and speed. "Pink" came a lone report from one of our riflemen and the air-plane pitched head downward out of control, crashing about two hundred yards away. O, how I wanted to go over for a look at "Heine," but the lieutenant now concluded there was safety under cover.

The appearance of the German visitor was not without its compensations. Instead of staying under cover at the village the men soon took to the streets and the wine-shops. I had been down the street seeking the quarters of some friend and while returning I passed Lieutenant Nysewander who was going in a great hurry across a vacant block headed for a noise coming from that direction. I took a short cut and reached the group of noisy men just before the arrival of the lieutenant. And there was King! What a sight to behold with only one pair of eyes!

King was joyously convivial. He had conceived the brilliant idea that he represented the entire Allied Forces. Being desirous of looking the part he had gone to a salvage dump where was stored the discarded clothing of all the men of the armies which had passed through the village. From it he had acquired an assortment consisting of one garment from each nationality. He wore a Frenchman's blouse, an Englishman's shirt, a pair of Scotch kilties, and his head was crowned with an American steel helmet unto which was fastened a red plume. He was 'strutting his stuff, saluting officers on every side. Lieutenant Nysewander, however, spied him and was bearing down upon him. Seeing impending catastrophe King called out to me: "Hell fire, Emmett. Look what's coming. Fini le guerre for poor old King." Drastic as was the lieutenant's discipline, he broke out laughing when he saw this regalia and said: "Corporal, get off the street and out of that garb,—before I see you,—or I'll be damned if I don't slay you with my own hands."

Automatic pistols were issued to the officers and non-coms here, and we were all anxious to try them out. We selected a site on the east side of the village for our practice but the firing soon resurrected another Boche plane. He forced us to conceal ourselves against the banks of a deep rock quarry. We did not have any long range guns and there was nothing we could do about his presence. Evidently he was not a bomber,—probably just an observer,—for he sailed over us time and again keeping us to cover but never firing upon us.[6]

They took us into the woods on a maneuver but our captain had been assigned duties as a major and was not along with the company. The trip

6 (Ed.) The first of two times that a lot of "Automatic pistols" were issued in the weeks and days before heading into combat. All the officers at the rank of captain and above in the Infantry battalions carried just a pistol as a sidearm. All the lieutenants who served as platoon leaders, and all the sergeants and corporals in the platoons and companies carried both a rifle and a pistol into combat. The automatic pistols were the US Model 1911 Automatic Colt.

was a hard one through the roughest country, through brambles, fences, briars, and almost impenetrable entanglements, but the men did not complain that night for they had seen a 'hard-boiled' colonel on a fine, prancing horse ride up to an acting major and administer a withering rebuke which reminded each man of the days back at Camp Travis when he was a rooky.

Chapter 8

Ready To Go In

News was waiting for us when we reached the village. We threw down our packs and discussed it with the K. Ps., who were left behind. Our kitchen had moved . . . and we knew what that signified. The front was only a short distance ahead now. Waiting for orders under such conditions might make one introspective, so Surrat and I started out on an exploratory expedition. We would see things instead of thinking. A French cafe was open and we purchased cheese and beer. We would not talk about what was going to happen, but one or the other would go to the door every few minutes and listen for the bugle. We knew it would blow soon. Finally, I arose from the table, counted my francs, exchanged my silver into paper, and stepped out into the street arm-in-arm with the big fat boy just as the first notes of the bugle called us to 'fall in'.

We made a run for quarters where everything was in a turmoil. Grabbing our packs we stepped into line to answer present. Darkness had come now. Trucks were again rumbling in the distance. L Company's billets were to the north of the town, and the column crunched along the road, now shoe-top deep in mud, . . . 'slop' . . . 'slop' . . . 'slop',—and we trailed into position. Fast moving trucks came up from our rear . . . "Give 'way to the right" . . . "Give 'way to the right", rang out the command in the darkness. We fell into the slushy pits along the road- side and let the heavy machines keep the road. What a night! A walk of a thousand yards and we halted. Trucks came after we had waited until we were exhausted trying to stand still. We loaded in with thirty-two men to the truck, each man taking his full equipment with him. There was just room enough for our feet on the floor. We had to stand with our packs on our backs for there was no room to place them elsewhere.

Corporal Ross, my diminutive friend, with his squad caught the same truck I did. I was the last man to load in, having remained on the ground pushing the other men up with their burdens. When my time came to get aboard the truck was full. Ross refused to pull me up. "No use to take excess baggage to the front anyway!" I swore at him, and Surrat gave me a pull. I fell on top of Ross and the fight started. The truck was moving down the road and I was trying to throw him overboard. "We'll see who is useless!" I would have pitched him out had I been big enough. We lunged around stepping on the other men's toes and they took a hand for self-protection . . . Just a little peaceable fight . . . no real harm done, a few skinned knuckles and a swollen lip.

Our engine developed trouble and we blocked the road. "Give 'way to the right, there" . . . and they pushed our truck into the ditch and we helped the driver start again. Now we had lost our guide and our place in the line. We had no instructions other than "to follow the truck ahead," but there was no truck ahead to be followed. Darkness . . . darkness . . . O, how dark it was.

All trucks were without light, and we depended wholly upon the night-eyes of the drivers. We entered a dense wood and felt our way along slowly . . . "Halt" . . . A man came from behind a tree . . . "Where the hell you think you're going?" . . . "To the front" . . . "Front, hell! You've arrived. The Dutchmen are right over there."

We managed to turn our truck around without drawing the fire of the enemy and retreated to a cross-road where we turned left to the bottom of a hill. At 3 a. m. we stopped in the darkest place in the imaginable world.

George Singleton had also been lost with a truck load of men, but an officer came to tell us we were to go 'on location' right now and lost men could stay lost men so far as he was concerned. Tramp . . . tramp . . . tramp, for a mile. Mud . . . slush . . . slop . . . and we held to the rifle of the man ahead and went on . . . on . . . on. A sharp turn to the left and we entered a trench, the clammy walls of which struck us about even with our shoulders . . . Mud was half knee deep. Rough stones were in the bottom.

We fell down in the muck and scrambled to our feet again to go on, now uphill, slipping, swearing under breath . . . climbing . . . always climbing . . . "Here it is . . . Spread out . . . Be quiet . . . Will see you in the morning" . . . (The guide was gone into the darkness and we never saw him again.)

Lieutenant Fly, now in command, decided a guard should be posted and I organized the guard. The men were tired. Clarence and I concluded to take guard-duty hour about, so as to let the men sleep. But ever faithful Surrat wanted to take his turn with us too.

The men lay upon the cold wet ground just outside the trench, huddling close to each other for warmth, while they slept. Then a strange sound came to our ears . . . "tinkle . . . tinkle . . . tinkle" . . . The peal of a bell floated to us from out in front. Sharp clanking of steel upon steel brought Surrat to his feet. With a stifled call of 'Gas' as his head went into his gas-mask, we were aroused from fitful dreams . . . We were hearing a front-line gas-alarm! All masks went on, but the inquisitive soon were sniffing the air again. Testing and smelling none, they came off again. Relaxing with the hope of not being disturbed again, the newly arrived troops fell to sleep only to be awakened again and again with the same alarm. And what a comfort was the first glimpse of returning day!

We found ourselves now lying under some very small bushes just outside a trench about four feet deep,—a deep gash in the earth extending over the hill and far down into the valley from whence we had come. Airplanes were hovering over the timber. We were told we were in the second line,—the first line of defense.

Stiff from cold, plastered with mud, tired from hunger and exertion, the outlook was not pleasing. Sergeant Allen saw me across a clearing in the timber and we approached each other. Then the air was rent with a terrific explosion. Allen fell to his knees . . . He laughed and looked at his helmet . . . A big dent was there . . . "Jerry has found my address!" And picking up a slug from the ground he said: "I think I'll keep this for a souvenir."

We now bedded down on the mushy earth like a bunch of hogs. Every man scraped together all the obtainable leaves with which to make his bed softer. So industrious was I that I became proud of my accomplishment. I was so tired and sleepy, and I thought of the "nightly mercy of the eventide" but an officer came along and commanded me to move over about thirty steps. O, the pleasure of army-life!

Regimental headquarters sent for a guard, and, of course, that duty fell to me. A guide directed me there where I found it had a semblance of protection. Deep rooms had been carved into the stony earth. Down there the 'brass-hats' took their comfort. There was a much deeper dug-out into which ran the telephone lines. Above ground and nearby was a cleverly constructed two-story building camouflaged so deftly as to make it hard to see. In this building the guards were quartered. By a peculiar coincident Joe Templeton had been ordered on guard with me. Hell broke loose in all its fury soon after our arrival. I had posted the guard and was making my second round of inspection when "Jerry opened up with both barrels." Heine was reaching for regimental headquarters and he filled the air with whining shrapnel. What a night![1]

I heard a moan and, thinking of dear old Joe, scrambled along the path edged with briars, which snatched at my legging at every step. I called but there was no answer from Joe. I called again. No sound came back to me except the 'bam! bam!' of exploding shells and the whistling zi . . . zi . . . zi . . . zoom . . . of speeding shrapnel. I concluded Joe had been killed ... I stumbled over a large stump. Wrapped around it was Joe. I pulled him to his feet . . . "Are you hit?" . . . "Hell no! Do you think I'm going to take it standing up in the dark."

"Crash . . . crash . . . crash . . . whew . . ew . . ew," and the shells came whining again. Then out of the darkness came the call of distress: "Corporal of the guard! My God, a doctor!" Passing rapidly in front of quarters and on in the direction of the calling voice I came upon three or four indistinct forms. Another was lying at their feet . . . "Is he hit, fellows?" . . . "Hit, hell! His leg's gone." . . . We tied a handkerchief around his leg stopping the blood the best we could. He told us his name. M Company and San Antonio had lost their first boy. They carried him off in the darkness . . . limp!

1 (Ed.) Both barrels of a double-barreled shotgun.

I went below to the telephone communication center which might be likened to the brain center of the human body. Perhaps Sir John Davies had a prophetic vision of these modern dug-outs when he wrote:

"Much like the subtle spider, which doth sit
In the middle of her web, which spreadeth wide:
If aught do touch the utmost thread of it,
She feels it instantly on every side."

Telephone calls were coming from all along the front. Orders were being given . . . "Line repairman for company . . . Communication broken by falling shell" . . . "Special order for all guards to watch for whip-poor-will." (A spy operating in our vicinity) . . . "Man killed at" . . . Thus the orders came in a constant flow. The soldier on duty at the switchboard was an agreeable youngster. I suppose he was lonesome. He was talkative and I was curious. Therefore, I soon had a head-phone set and was plugged into the switchboard. An interesting panorama unfolded to my ears.

I was relieved from duty at four in the morning and sought out a bunk in the second story of the guard-house where I soon fell asleep. German gunners, however, would not have it so. They began pouring high explosives into the sector again. A terrific explosion came directly above me, and a part of the roof over my head flew away. Following closely behind the first one, another struck within a few feet of the first,—uncomfortably close to me. A pale daylight came through the hole to warn me the night was over. Then I heard a noise recognizable as the familiar tinkle of mess-kits and sleep no longer was a necessity. I filled with 'carrots and coffee' and launched out into the vicinity to see the sights. Nearby was a road over which a vast number of men and great quantities of military supplies were moving. Travel on the road was deftly obscured from the enemy by high trees and man-constructed camouflage fences. The road branched, one part going to the right and behind a precipitous hill, while the other took the course to the left and westward passing up a valley far up which our company had scrambled in the darkness a few nights before. The timber was very dense, affording fine concealment for troops.

Morning passed quietly with little shelling, even though the new troops were indiscreetly showing themselves from under cover in many localities. Dinner-time found me far down the road keeping close to the breast of a bluff for protection against the scattered shells. Here I walked upon dear old Rogers Tipton who was answering his internal-chow call. We filled our messkits with 'monkey-meat' at a nearby kitchen and retreated to a point near the embankment to sit upon a log to enjoy what was to be our last dinner together.

I was greatly shocked when I first saw Rogers that day. He had left Jacksonville with hair completely black. That day his hair was almost completely grey. I chided him about getting "scared white" but he seemed not to be concerned. 'Jerry' showed us he knew our movements by throwing a salvo of 'heavy ones' close up to the kitchen, after which, failing to register a hit, he turned his mind to other targets. We finished our dinner and I started back to headquarters. Tip called after me: "Won't be long now."

The approach of night brought my relief and I wended my way toward the company sector. As I came down a trench I was singing:

"Just a song at twilight,
When the lights are low,
When the flickering shadows,
Softly come and go:
Though the heart be weary,
And the days are long,
Still to us at twilight,
Comes love's old sweet song."

I heard the noise of approaching feet, but knowing I was within the company area I continued with the song. Bill Thompson presented himself squarely before me from around a jog in the trench. Tears were running down his face. He pulled his bayonet and threatening me with it, said: "Damn your honorless soul, if you sing another word of that I'll cut your head off smooth with your shoulders." . . . I should have known better for Bill had been married shortly before leaving Texas . . . and he was in the throes of recollections!

Darkness was closing around us rapidly there in the timber, and with it a general order was put into effect that all men should "stand to",—meaning that every soldier should occupy the trenches and stand in readiness for an attack. It had leaked over to the Germans, so the rumor went, that a new American Division was occupying the area and a night attack was going to be launched to test American preparedness.

I took my place in a deep cut and waited. I spied something out in the semi-darkness and being curious, decided to investigate. Advising my comrades of my intentions, I left the trench and crawled out. I found a German carcass. Lying nearby was a machinegun with a tell-tale hole through the magazine. Spread out from the gun was a partly fired belt of ammunition. Here was an unrecorded story.

Our wait was long. I had had little sleep since we came to the front. Human strength could no longer endure such trials and I leaned against the parapet sleeping fitfully. The slightest unusual noise, however, was sufficient to bring me again to the realities. A tread on a rough stone would have awakened me. The consequences of sleeping on guard-duty were deeply ingrained into our minds . . . but we were human.

A few days passed . . . I have no way of knowing how many . . . Time began to roll over us in such a way that we could not tell whether today was yesterday or tomorrow. Time meant nothing . . . We were facing physical realities . . . rainy days . . . wet to the skin . . . The cold entered our souls . . . There was cold food . . . luke warm coffee . . . and time. Time, of all the hardships, was the hardest to endure. As the darkness of each succeeding night settled around us Jerry would heighten our spirits with a serenade of long range guns. A shot would dab here . . . another would strike there . . . and there . . . "boom; . . . boom! . . . boom!" We set our watches with their accuracy. Just a parting "Good night."

Again at ten another salvo would split the night. All up and down the front the shells would whine. After one of these bombardments a terrible rapid firing came over to us. "Pop . . . pop . . . pop" . . . myriads of rifle explosions rushed us to a 'stand to' formation. We knew the long expected attack was coming! After standing ready for action for an interminable time we were informed that a big shell had made a direct hit upon an ammunition dump belonging to the division and the shells were burning. The noise continued throughout the night.

Disgusted with inactivity, uncomfortable, and chafing under restraint, the prospects of an added venture now became most welcome to all of us. And the order came just as darkness shed its gloom over the woods. It was: "Move out rapidly. Hold onto the gun ahead of you. Speak to no one." In no other way, however, would it have been possible for us to have moved, for contact with the man ahead could only be maintained by holding to his gun. The darkness became too intense to see even the outline of a man ahead. In this manner we passed down into the valley and were crossing over to the right of the road when a team ran away in the darkness. I was struck by a passing mule and knocked off the roadway. Not being hurt, I arose and tried to get back with my company. This was not the simplest of accomplishments, for the entire division was moving up that road. The road was absolutely blocked with a mass of moving equipment and men. Loss of 'contact with the gun ahead' became a serious matter. I caught someone passing and asked who it was, hoping to establish the proper contact by the inquiry. I got no response. I asked again and got the same results. The din of moving horses, mules, cannon, men, everything imaginable . . . tractors chugging and snorting . . . drivers swearing . . . sergeants calling to their men . . . made me know that the general order 'to speak to no one' came under the caption of 'just another one of those orders'.

Then and there I determined to have an answer or know the reason why. I grabbed the man next to me by the throat and shaking him until his eyes must have started from their sockets, demanded in tones firm and positive: "Now damn you talk. What outfit?" Imagine my surprise when he said very calmly: "Lieutenant Fly speaking, Corporal Emmett . . .

L Company 359." I side-stepped and ran into other soldiers and found they were a part of my company. I trailed along until I located my squad . . . and the long walk began.

To one who has not seen and heard it there is no way to conceive just what it means to have a division of soldiers on the march in the night time. Horses and motor-drawn vehicles must, by the very nature of things, occupy the roadways. Officers going to and fro to give orders must take the roads. There is nothing for the poor infantryman to do but "Give 'Way to the right" while the thousands of cannon, more thousands of horses, and chucking motors push onward. The clank and clangor of metal, grunting, puffing motors, loud cursing soldiers, the interminable rhythmic tramp, tramp of hobnailed shoes raise a din beyond comprehension.

When once we moved into the line of march and assumed our proper position many of the difficulties had passed. No longer did we 'scramble up' with run-away teams nor collide with a stinking, clucking tractor. But the steady pat of feet resounded throughout the night and we had a feeling of great strength. We were a great force flowing, throbbing with power. I was a part of this great power and I was glad I came along.

We pressed onward now, and lurid streaks of light pushed up in the distance. Then followed a trembling of the earth. A growling deep-throated 'boom' slashed us back to consciousness . . . "There lies Germany."

There were no songs that night . . . just a steady grind of feet striking the stony, sloppy road. There were no bantering shouts to your buddy that night . . . just a whispered anxious inquiry: "How're yeh comin', Bill, OK?" And a sincere tender of assistance to your bunkie: "Let me carry your pack, old top" . . . And a word of encouragement: "Stay with her pal; can't be long now." . . .

Argie Garner had been walking by my side during the grueling grind of this never-to-be-forgotten-night. He struggled . . . then lagged . . . disappearing into the darkness behind. Some minutes later he appeared again at my side, running, out of breath . . . He had gone to sleep while walking.

We wound out of the timber and spied the outlines of a camouflage road above us . . . The skyline grew lighter . . . a hazy vision . . . a white road . . .

soldiers . . . soldiers . . . soldiers stretched in one indistinct, unbroken line as far as we could force our vision into the night. A turn to the right, a long walk around the rim of a hill, a left turn, and we trudged out onto the open prairie. "Slog . . . slog . . . slog" we punctured the night with our motion. Flash after flash came near in front now . . . and no one spoke![2]

A great black blanket covered the earth ahead,—a wood. We came to a cross-road. A stone-house with the whole front caved in stood to our left. It was obscured by the timber to its rear. We halted. O, what a relief . . . aching shoulders . . . burning feet . . . and we collapsed in the mud to the right of the road . . . "Whee . . whee . . ew" . . . and a ripping scream rent the air above. Louder . . . louder . . . it came . . . Bam!! Light and sound seemed to have been born together up there over our heads . . . and shrapnel fell all around us. 'Jerry' was shelling the cross-road, and we—green troops—had selected this point to rest. Cries of pain . . . whimpers . . . commands: "Move out of the road . . . Over to the right . . . under cover . . . the trees", excitedly came to us. We dragged weary bodies aside, reanimated with fear, and fell down again.[3]

Jerry seemed to have satisfied himself with one salvo, for he moved his range and the shells began to fall in the timber ahead of us. There was a long whine, a flash. Screams from pain-stricken, shrieking horses mortally hit brought us to our feet. What terrifying sounds a stricken horse can make! But all was quiet again. The night settled down to a steady pink flash . . . flash . . . of guns far away, just puffs of reddening glow with no sounds.

An officer passed quietly along the groups of prostrate soldiers . . . "Move out quietly, boys. You are taking over the first-line trenches . . . Watch out for raiders." We struggled to our feet to enter a forest of impenetrable darkness. Our course was down-hill. We made a sharp turn to the left, leaving the road, and snaring ourselves on briars while we struggled through the churned-up mud. Now we struck the 'duck-boards'

2 (Ed.) To hide the movements of troops and vehicles from enemy observers on the ground and in observation balloons, cloth and rope netting was stretched over exposed portions of roads near the front.

3 (Ed.) Crossroads were particularly dangerous places to go through. Because of traffic, vehicles and troops often slowed or stopped for traffic from both directions. The junction was easy to locate on maps and from a distance. Forward observers could often hear the grinding of gears, the press of brakes, or the sound of engines accelerating in the approaches and call for planned fire to catch the tail end of convoys before all the vehicles passed.

already slick from the slime of many feet . . . We stumbled . . . shambled . . . and forced our way along. The darkness had become intense . . . The overhanging trees precluded the passage of a vestige of light. I fell, and went off the duck-boards. I was literally encased with mud . . . "Take my gun, soldier", and with a heave a buddy put me back into line. We struck a wire fence . . . and the sharp barbs sliced into our hands . . . We followed it along . . . Dark forms sat on the ground. We knew we were relieving them . . . Not a word was spoken . . . We came to a clearing, and after a quarter-mile walk we entered another strip of timber. An officer whispered: "Turn left . . . Down . . . down into the trench." Down I went to find I stood in water, mud, and slime. I felt around. I found the wall. It came up to my shoulders. Another officer was whispering: "Here . . . stand-to . . . Expect a raid, now, any moment . . . Shoot anything that moves."

I lifted the heavy pack from my shoulders and plunked it down into the ooze. It fell with a squashly thud. My legs ached. I tried to loosen my leggings. My hands went into the slime. I was standing in muck over my shoes. I checked-up on my squad. All were in their places, and we sat down in the mud, for we didn't have any other place to sit. A chill dampness crept up our bodies and we stood up again. It began to rain. We found a trench-pick and began to dig into the side of the trench. We tried to make benches . . . anything to get out of the cold drizzle falling on us . . . I finished a hole in the wall and crawled into it. It was some improvement. Others were doing the same. And there was a general noise all along the trench,—men working,—digging in!

We did not have long to wait. Far out to our right-center came a whine . . . louder and louder . . . and then the crash. The shell broke over-head and the shrapnel fell about us. Another and another followed. Daylight began to make its appearance. Jerry's alarm-clock, true to form, was waking up the entire front![4]

Hazy forms began to take shape. Tall timber showed up in front of us.

4 (Ed.) "Shrapnel" was a common small artillery shell. Rather than a full explosive charge, the shell was filled with numerous half inch or so round lead balls. The time fuse was set to explode the charge in the rear of the shell to throw the balls downwards in flight like a huge shotgun blast. These shells were designed to cause casualties among men or horses. It was the reason all armies developed steel helmets during World War I to protect Soldiers from shell splinters and shrapnel balls raining down into their trenches.

We were on the very outskirt of the wood. Back of us was the dense forest. We had walked almost through it in the darkness. As it grew lighter we could determine our position more clearly. We were new men, inquisitive, desirous of knowing what was before us, and carelessness was the natural result. A hissing sound sped past me . . . again and again . . . "Pst . . . Pst . . . Pst" . . . It was a new sound to me. A sharp cracking noise floated to my ears,—rifle-fire! Germany was near! Inquisitive heads withdrew below the parapet. Mine was one of them.

All down the trench our men were in motion, walking with stooped shoulders . . . keeping heads below the level of the ground. Officers came and went . . . "Heads down . . . Don't show your position!" The officers went away, and heads again were thrust above . . . Away out in front there was motion. I watched. What was it? Motion again, —a flag . . . a black-flag! The symbol of death! . . . No quarters! "Damn nervy devils, eh"? said a calm old boy, as he wiped the back of his dirty hand across a stubbly, tobacco-stained chin. Then he looked carefully at his rifle. It was loaded. He walked along the trench and stopped at a point behind a tree. Over the parapet he went. He crawled on his belly, slowly, carefully. He lay quietly. I watched with bated breath. He was aiming. "Crack" and the black-flag tumbled over! He crawled back to me, and as I helped him down again, he said: "Guess that'll show 'em who's doin' the shootin' 'round here, eh?" . . . Down the trench an animated whispered conversation was taking place, and a placard arose from our trench in full view of the Germans:

"Give your hearts to Jesus:
Your tails belong to us."

There was a volley from the German lines, but the placard still stood.

A light wind came from out the north. Rain ceased to fall, but all of us were wet and the wind pierced us to the bone. A lazy sun tried unsuccessfully to peer at us through a murky sky. We sank down in the mud, waiting, waiting, always waiting . . . and for what? Food . . . action . . . anything . . . just so we could move!

A tinkle of mess-kits brought important news. In groups we moved toward an obscuring, heavy undergrowth where stood two of our faithful boys. Two G. I. cans set on the ground. One contained coffee, the other 'slum-gullion.' I came abreast and held aloft my tins. Old Chloupek, that lean, long, lanky, inert friend-to-everybody, smiled at me: "Sorry. Not much this morning. Hard to make a fire up here without Jerry blowing it out". I slopped back to my location and lifted the cup to my lips to drink. The coffee had a thin skim of ice. I broke the feathery patterns with my fingers and drained the cup.[5]

We spent the day in quietude but moved about occasionally to start the circulation. We were always polishing the old rifle. Rub . . . rub . . . rub . . . That helped some . . . Anything for motion.

The sky cleared a little and the mounting sun cast its rays 'out there' lighting up the ground. I had an impulse to see, so I left my position and found a point of vantage where I could overlook a wide area of the sector. We were 'dug in' at the very timber-edge. The wood ran south for about three-quarters of a mile, but a short distance north of my location the trend was toward the north-east. Our position was barely within the timber-line and we occupied the nearest point to the Germans. Out in front, and to the left, was a little vale which deepened as it descended toward the north and east. Across this valley was a veritable bramble of tangled wire-fences. Row after row these brambles stood as silent protectors before occupied German trenches. Set well back into these trenches and slightly over the crest of the rising land stood a small group of houses. It was Fey-en-Haye. Looking sharp to my right and following the valley before our trench, I saw the steep walls of a gulch pock-marked with dug-outs. Only occasionally did I see the protruding steel point of a German helmet or the bucket-like hat. No sign of life showed at the hillside dug-outs. I looked long and intently trying to determine the cause. Then I could see that our rifles could reach the dug-outs, hence, they had been deserted. The Germans had taken to the trenches on the brow of the hill on a level with us so they

5 (Ed.) The G. I. cans were General Issue cans. Hot food and coffee were often brought up to the front in large insulated cans with tightly sealed lids known as Mermite cans. While the shape and design have changed over the years, the U. S. military containers to transport hot or cold food to troops in the field have the same name.

would not be disturbed within their abodes by our rifle fire.

Evening came and my old friend, Ross, with his squad of men moved over near my location. A trench which ran deep into the woods, extending from our trench, intersected almost at right angles within a few feet of my position. Ross decided to move into that sector. I discussed with him the danger of getting where he would not be protected by a sidewall, but disregarding the admonition he lay down against the south-east wall of the lateral. Just back of me and just in the fore of Ross was a tree about thirty feet tall having a diameter of some fifteen inches. Ten o'clock came . . . the time for the change of the German gunners, we learned later . . . and the unfailing salvo came over. They must have recalled the fall of the black flag for shells came all along the line. One, two, three, four, invariably they came four in a place. ONE . . . Bang! High in the air above me. TWO . . . A little lower, but too high to do any harm . . . THREE . . . A long whine went across the tops of the trees exploding far back . . . FOUR . . . (I counted.) A whine . . . Bang! A terrible concussion. Tree branches fell all over me. I looked up and the tree was falling . . . toward Ross . . . It struck across the trench burying him in its down stretching branches. A frightened cry for help came from out the leaves. I went to him. He was lying against the wall, pinned there by the Branches. I yanked at him . . . pulled him . . . and he came out, spluttering . . . spitting mud and leaves . . . and unhurt . . . But he stayed with me the remainder of the night! When morning came we inspected the damage. The tree had been struck about ten feet from the ground and it was clipped off as neatly as if done with a gigantic paring knife.

When we became quiet again after the shelling subsided an officer came along the trench and gave us the password, "Austin" and he took away my old fat friend, Surrat, who was to go on a night-exploratory expedition.

Poor old Surrat wasn't so enthusiastic about this first raid on the German trenches . . . He handed me his watch, his note-book, even his francs, and admonished me to give them to his sister if he didn't come back, and . . . sentimentally . . . I promised.

Their order was to proceed until they made contact with the Germans, then, estimate the distance and retreat. This was done, according to later developments, to ascertain the straight-ahead distance to the German positions so a barrage could be laid down on them.

Surrat had not been gone long until the staccato 'tat . . . tat . . . tat' of the invisible blacksmith hammering away in the darkness disrupted the quietude of the inky night. They had, evidently, made 'contact' without difficulty. Surrat arrived shortly thereafter.. He was all aglow with enthusiasm. His fear had left him. He was telling what fun it was . . . I ought to try it . . . I did not agree with him!

The night wore on, and morning came. Jerry shot at us as usual. There was no harm done near me . . . and we were served coffee and slum-gullion . . . and we went back into the trenches again. I had had no sleep since we came here and my brain was getting groggy, so I wrapped up in my blankets, my over-coat, and covering with my slicker crawled into the manhole in the sidewall . . . But I could not sleep. The sky remained over-hung all day without a sign of the previous day's encouraging sun. My feet ached . . . Then they became numb . . . I took off a shoe to rub life back to my toes. A slimy silt had ground itself to the very skin of my feet . . . Rubbing did little good, and I got up and walked about. Evening's approach threw a murky gloom over the woods. Other men stirred around . . . They were restless . . . careless . . . More rifle fire . . . spit for spat . . . Officers passed along the line complaining we were making the Germans 'unnecessarily nervous' . . . "What's the diff? . . . Didn't we come to fight . . . Then, let's fight . . . not freeze to death in a mucky rathole" . . . And two meals a day now . . . "Two much trouble to carry the food so far" . . . "Yeh?"[6]

An officer entered the trench to speak to a sergeant. The word 'came down the line' . . . "Stand to with fixed bayonets all night tonight . . . The Germans will send their raiders over in retaliation" . . . "That's better anyway . . . Something to do" . . . But throughout the long weary night we stood with heads above the ground, standing first on one foot then shifting in pain to the other . . . Ten o'clock . . . and the usual heavy cannonade . . .

6 (Ed.) While the term slum-gullion has many origins, the Doughboys were referring to a form of "refrigerator stew" that company cooks produced out of whatever they had available at the front in the quantity required to fill the mess kits open in the line leading to their mess wagon. The A. E. F. began to issue ever larger amounts of canned food that was transported across the Atlantic from American farms and canneries. Cans full of stewed tomatoes, corned beef, corned beef hash, diced potatoes, any onions found, any stray canned vegetables, and sometimes a few cans of salmon, were added to the pot to make the variable "SLUM" of the Doughboys.

Then night slipped gently into breaking day . . . and the cannon boomed again. Exhausted . . . and disappointed . . . we faced another gloomy day . . . The rain fell and we walked about in the trenches . . . slop . . . slop . . . slop. "What was the use? We sit in mud and shiver. We stand all night straining our eyes against darkness . . . and no enemy appears." Sardonically I warned Surrat to 'keep his head down . . . He might get shot'. He wants to know why I have an interest in his head when he would give the Crown Prince himself three cracks at it for a pull at a bottle. Then when Surrat speaks of the bottle I think of the kitchen and become desperate. I walk over the rear of the trench and step into the woods. I do not know where I am going and, furthermore, I do not care. I saw some men about one-half mile away. I went toward them and discovered it was our kitchen. The cook seemed to be glad to see me and smiled while he started after a cup of hot coffee for me. Jerry threw a big one down in that direction about that time and he jumped down into a little dug-out the K. Ps. had made, and I got my own coffee. Captain Whitaker came up about that time . . . "How's everything, corporal? . . . Out under orders? . . . No? . . . Well, be careful . . . and say I sent for you!" . . . "Not such a bad fellow," I thought, "after all." . . . but when a man has plenty of hot coffee the world does seem different.

I walked slowly along a skirt of timber keeping well under cover. I would go back to the boys. I was sorry I had drunk the coffee now that they could not have some too. I came to a trench filled with men from my company. They were talking . . . They wanted to tell me about it also. One of our men had fallen into the trench the night before and had claimed to have injured his back . . . My mind went back to years of previous experiences . . . Some one said: "Damn yellow quitter" . . . Perhaps? . . . But I could not keep from thinking: If he was determined to quit this war, well, right now was the time to make the decision . . . and he had gone to the hospital.

I trudged slowly onward and it began to dawn upon me that there were only three kinds of soldiers,—buddies, those who stayed with the company regardless,—'lucky-devils', slightly injured boys, sufficiently to be 'sent back' (Blighties, as the English called them),—and "the damn yellow quitters",—who . . . but why explain!

These classifications had taken permanent form in the minds of the soldiers. There was universal admiration for the 'lucky-devil', but for one whose cause for being 'sent back' was not apparent and outstanding,—an actual visible injury,—there crept sedulously into the minds of his comrades the idea that he was a 'damn yellow quitter'. No man, therefore, even when obsessed with fear, who retained or wished to retain the respect of his comrades, ever thought of the blessings of an 'invisible blighty'. If he did, he never spoke of it even to his closest comrade. Perhaps he might wish upon himself sudden and violent destruction but there was a ban even on the thoughts of an invisible blighty . . . for fear of the opinions of his comrades . . .

Many and grievous injustices to reputations of men were visited upon soldiers by the 'damn yellow quitter' classification. Men would persist in going on, and on, and on, until racked in body and mind by the hardships of war, fearing the cruelty of the condemnation of their comrades more than pain. They preferred exhaustion to bringing their reputations into question before the eyes of their stronger comrades.

"Damn yellow quitters", however, did exist, and there were some 'lucky-devils'. There should have been more of the latter. Had there been, then the aftermath of the war would not have registered so many shell-shock cases.

Skulking along, these ideas were forming in my mind. I, therefore, became impressed with the necessity of immunizing myself from the effects of the pangs of physical endurance. I saw that physical pains had their compensations. Stark tragedy parading over the corpse of civilization, without the compensating distractions of physical pains, would have turned the most stoical of us into opportunists. Soldierly obedience would have crumbled . . . I determined to accept the days as they came, for . . . what was the difference, anyway?

I reentered our trench at the point where we first went down into that stench-pot some nights before . . . I looked . . . and I laughed. Sitting in the mud, posteriorly encased about six inches high with bubbles oozing up around him, was old 'Red' Crow. A blanket had been folded over the mud. Kneeling before Crow was the Mexican boy, Florez. They were settling their differences in a crap-game.

I found my squad again and relayed the news to Oliver and Surrat. Then I hastened on with bended back,—so as to obscure my head,—until I came upon old King, who, true to his style, had dug into the side of the wall. Ganged around him were his 'fighting hellions' to whom he was relating the minutest detail of his latest lurid experience, born out of the fervency of his imagination. He stopped as I appeared. "The empty vessel makes the greatest sound", I warned him. But he scooped up a hand-full of mud and spattered it on me.

The day dragged through and I was called to the P. C. Night-guard duty fell to me, and as I was corporal of the guard I had to keep contact, post to post, through-out the night. In order that this might be done, of course, I had to go out when the guard was posted, find the locations and remember them so I could return without the aid of anything except 'recollection' when night came.[7]

The guards were posted along in front of and near the trench occupied by us. At one point our trench ran out into an inverted 'V', the point extending toward the German lines. We stationed three men here. Two were in the bushes in the fore of the triangle and the third man, good natured Pete Jones, was placed to the rear of the advance-point men at the intersection of the two trenches.

Darkness came with an unusual intenseness. Hour after hour I felt my way along from post to post. Not having closed my eyes in sleep since we entered the trenches I was 'blind' tired. I could make my rounds leisurely for everything had gone off free of excitement. About midnight, after having been to Jones at his out- post and having learned of nothing to disturb us, I went into the main trench and proceeded to the most northerly extension. Suddenly there burst forth a spitting which I concluded was machine-gun fire. Seeing the flashes and recognizing the location as the advance-out-post, I concluded the Germans had brought up a machine-gun with which

7 (Ed.) The P. C. was the abbreviation for the French term "Post of Command." The British term "Command Post" became more common in the United States military after World War I and since.

to mow down our outpost. I hurried to the spot. I called within the hailing distance, but received no reply. Repeatedly I gave the call but received no response. I advanced to Jones' position where I found him hugging the intersection of the trenches, too scared to talk coherently for a moment. He then explained that no one except himself had fired. As the fire had been so rapid I could not believe him. He contended, however, that he had been "almost in a doze" when someone had fallen into the trench just to his back. Then he had "held the trigger down and pumped her dry". It was the fastest shooting ever done with a bolt rifle!

We searched around trying to find a dead German. There was none. We did find, however, a German rifle, a steel helmet belonging to an enemy, and a 'potato-masher', unexploded.[8]

It was plain the German had fallen into the trench almost at the feet of Jones after he had attempted to circle back and approach Jones from the rear. His attempt had only failed because he had accidentally fallen into the trench . . . where he was shot 'at' by Jones. I took the German accoutrements with me to the P. C. where, I suppose, they were appropriated by some officer as his prizes,—a very common practice by those in authority . . . not that it amounted to anything!

There was no chance for us on guard to get any sleep that night. We would have enjoyed the opportunity to sleep next morning, but as the sky cleared a little, the temptation was too great to sit under the warming rays of the sun, permitting our soggy clothing to dry a little. We burnished up our guns, and finding a convenient crap- game in progress on top of the mud we took turns at the blanket. Coffee came a little late that morning, but it came warmer than usual and without the skim of ice. We were beginning to think we were to have a break with Lady Luck, but . . .

Shortly after the arrival of the coffee I received orders to report to the P. C. which I found about one hundred yards back in a dug-out. My arrival there caused me to register a thought of some importance! The order was for a number of non-commissioned officers to accompany an officer on a day-patrol into 'No Man's Land.' It was particularly pointed out

8 (Ed.) "Potato masher" was the slang term for the German defensive hand grenade used in both world wars with an extended wooden handle to aid in throwing that made it appear similar to a kitchen utensil.

to us that we were not to arouse the 'ire' of the Germans but we were to familiarize ourselves with the entire trench-system lying between us and the enemy. No one expressed a desire to start a 'one-man's-battle'. All of us were very thankful to have advantage of the little previous information which had been gathered by the night patrols indicating about how far we might expect to proceed with reasonable safety before establishing contact with the Germans. There was no way, however, for us to know whether or not the enemy had established intervening outposts. With this we had to take our luck. We had special emphasis laid upon the necessity of our making a careful study of the trenches. Admonished to make no notes but to get the situation so thoroughly in our minds that we could return to the area without the aid of the light of day, the question became no longer a mooted one with me, for I saw we were soon to occupy that territory. An attempt to occupy 'No Man's Land' meant a fight, and a fight meant a drive along the whole front. We adopted those conclusions immediately upon receiving our instructions. The Germans, we knew, were entirely too well 'dug in' to allow us to take over such advantageous territory short of a finished fight for the advance.

Our instructions from the officers concluded with an admonition to be ready to advance into No Man's Land within 'the next few minutes'. It was now my time to think of the consequences of that reconnoiter. I chuckled when I thought how scared Surrat had been; now I was worse frightened than he. I had, however, little opportunity to think, for by the time I had returned to my position a messenger came with instructions to join the other men.

We took a course northward until we came to the side of the cliff before the valley. Here I was surprised to find the trench running down the hill completely covered with wire-netting and grass. This camouflage effectually obscured us against peering observers, and we went across the valley entirely under cover. The camouflage-veil was thin enough to permit us to see through it, and we had a good view of the lines of trenches running across the valley and their entry into the hill beyond. Our progress was slow and attended with the greatest care. Each man made all possible observations. We realized great consequences might hinge upon the accuracy of our observations.

After filing down the hill-side and through the valley under camouflage protection we came from under the obscuring veil and began the climb to the heights beyond. With the hill before us there was no possibility of the enemy seeing us from their trenches. We could have been seen, however, from passing planes. The fire of the boys from the trenches was forbidding to the air-men, and we were fortunate that afternoon, for none came our way. As we climbed we found the trenches to be very deep, sufficiently so to permit our walking erect without disclosing our positions. When once the crest of the hill was reached we intersected a perfect maze of trenches. These were deep along the crest but the farther westward they extended the shallower they became. Once out from the brink of the hill a hundred yards, our heads were showing above the level of the ground. This made us go in a stooped posture, a position which became very tiresome, especially as we had to continue nearly one-half a day.

Our general course was north by west. For weapons of defense we had taken rifles, pistols, and hand-grenades. These made a burdensome load. I became exhausted trying to carry my burden and proceed with caution. I came to an intersection of two trenches, one pointing southwest and straight toward the little village of Fey-en-Haye, then in the possession of the enemy. I stood quietly peering from around a protecting jog for a long time, trying to locate some German activity. Evidently the enemy knew the town was under observation for none showed themselves upon the streets. While thus looking, however, I did find something which bristled with possibilities. Not very far from me, a German with rifle carried straight up, bayonet fixed, passed across the lateral down which I was looking. He did not see me. With his head sunk over on his chest, he moved out of sight. Strange as it may seem, I had the greatest urge to shoot that man. It would have been the easiest thing possible for me to have killed him for he strode along leisurely, wholly unaware of the near presence of anyone. This man,—the first evidence of an enemy being near us,—indicated how close we had crawled to their first-line.

The terrain from this point onward took on a different contour. It was slightly down grade toward the Germans. The enemy had barricaded themselves against incursions by cramming the trenches with barbed wire. In some places there was some semblance of order in their defensive use of wire, but in others the protecting trenches were literally piled with rolls of the ensnaring barbs. At no place was it possible to proceed the full length of a trench without encountering an impassable entanglement. This created a serious situation for us as our officer insisted we should complete the survey. To do so required that we leave the trenches every few yards. So, on we went. Finding an impassable wire-entanglement we lay upon our bellies and slid with imperceptible motion out into the open with nothing between us and the Germans for protection. The first time I eased myself over the parapet and sprawled flat out on the ground to begin what seemed to be an interminable crawl ahead, momentarily expecting to feel the sting of a bullet, my heart cavorted uncontrollably.

It is a known fact that motion is what really attracts the eye. An object may be in plain view and it is seldom observed unless it moves, or unless there is too great a contrast in color. Therefore, proceeding on this theory, our progress around the wire-brambles, from trench to trench in the open, was steady and imperceptibly slow. We moved, not inch by inch, but with an unfaltering drag, pressing close to the ground, with face turned from the enemy, and one man at a time. There could be no flash of the hand nor jerk of the foot. No sheen of light off Caucasian skin should warn them. There must be only a steady crawl much after the manner of the slither of the snake.

On and on we went throughout the evening . . . Seldom was a word spoken. If so, it was a mere faint whisper,—a communication more by lip-reading than voice. Finally we came together in a depression about shoulder-deep. Here we rested while the officer looked over the top of the parapet with a trench periscope . . . Then each took his turn in looking . . . Just beyond us were the tips of moving bayonets and an occasional steel-helmet.

The officer now agreed that to go farther would accomplish nothing beyond being discovered, so we started our retreat. By this time the evening had almost gone, and we did not want to be out there when darkness came. We could hardly restrain ourselves from fleeing toward the protection of our buddies now so far away. Our success, so far, had been complete. To hasten now would mean discovery. Our efforts would have been ruined. And we steadied our nerves and started backwards, with an even greater stealth but always with an urge to look toward the rear to see if someone had discovered us.

When we reached the rim of the protecting hill I ached in every muscle. Fatigue, lack of sleep, and nervous tension had so completely possessed me that I could only with difficulty collect myself sufficiently to walk down the hill. Hours before we had climbed that hill with apprehension. This same hill, now, to us was a mental palladium. There should have been a pleasurable exhilaration and satisfaction culminating our return, but strange, indeed, this was not my impression. A look at the strained faces of my comrades showed that they, too, registered none of the ecstatic. We could not shake off the pall of the afternoon. I did not seem to care what was to come next. I could not enthuse over the accomplishment.

I mustered enough energy to walk across the valley, which was now black from the shadow stretching out from the hill. As I entered the covered trench to climb back to our position, the sun—just for a moment—flitted a red glare through our covering . . . and I thought it was symbolic of blood. Night began to draw around us. I squashed listlessly along to our squalid quarters . . . I dropped down in the mud . . . It had been a hell of an evening!

At first I thought I would not respond to the tinkle of mess-kits, but the slop, slop, slop of trudging feet and the ever present pangs of hunger aroused me . . . I found myself drinking cold coffee . . . eating slum-gullion . . . laughing with old Surrat . . . "Well, I know where they are, too!"

Darkness enveloped us. Jerry saluted us with his usual 'four guns'. There was a perceptible nervous tension all along the front. Everybody sensed it. Jerry was extra-ordinarily busy with his machine guns . . . "Tat . . . tat . . . tat" . . . the sound came floating to us from a dozen new positions. If there was ever an atmosphere which 'impended' possibilities, ours was such that night. The American lines were deathly quiet. We sat in the mud and listened while nervous rigors raced up and down our spines. The usual 'complimentary' exchange of big-gun salvoes did not occur. Heine simply 'tested for our positions' and subsided with no response!

Chapter 9

The Battle of Saint Mihiel

A hastening footstep came along the trench. Ward, our bugler, now turned captain's messenger, for we didn't need a bugler, approached: "Corporal Emmett, . . . The captain's orders . . . Report to the P. C." . . . He was gone.

Back toward the rear, blinded with darkness, a sludging stream of non-commissioned officers collided with each other as they felt along the guiding walls. Violent oaths . . . whimpering half-undertones . . . and 'slop' . . . 'slop' . . . I followed the sounds . . . Three trenches came together. At their intersection stood a soldier . . . "Non-com?" . . . "Yes" . . . "L Company?" . . . "Yes" . . . "To the right . . . and watch for the steps . . . They're slick" . . . and I ran point-blank into something soft . . . I felt it,—a blanket stretched over a door-way. I pulled it aside, proceeding. "Careful there, soldier." And I found another blanket, a screen for the light in the dug-out . . . I pulled it aside . . .

Pale candle-lights flickered weakly from their place on top a steel-helmet. The commander sat in the corner . . . on the ground . . . Several officers . . . I could not recall having seen all of them. Again and again the blanket was flicked aside and non-commissioned officers arrived. The captain was speaking to me: "How're you making out, corporal?" . . . "O.K., Sir." . . . "Had any sleep?" . . . "Lots of it, Captain, . . . that is, . . . back in America . . . Don't recall arry since we moved in here." . . . "I thought so." . . . Another messenger arrived. He handed something to the captain . . . Nervously he fanned the paper against the fingers of his other hand . . . "It's to be at eleven tonight, men. . . . Move out at eleven . . . Over the top at five . . ."

There was a depressing stillness. My ears were ringing, but there was no sound coming to us down there in that dank place. There was mud on the floor of the dugout brought in there on the boots of the men. The blanket over the entrance was still now since the men were not coming and going, and the candles stuck atop the steel-helmets sent their yellow flames upward without a flicker. I could smell the burning paraffin . . . and the odors of men . . . And for a long time, . . . so it seemed to me . . . no one spoke . . . Then from over in the corner an officer said: "Sir; these are the men who scouted out the terrain this afternoon."

It is singular what observations a man will make under stress of circumstances. I was no longer impressed with the approach of the hours of eleven and five. I could hear, but the sound seemed to come to me faintly from a great distance. I knew that plans of attack were being discussed . . . "Move out at eleven . . . Attack all along the line at five." . . . but before me . . . and how funny it seemed? . . . I wanted to laugh! . . . sat officers,—not the trim, neat men I had been saluting, but men with whiskers . . . men with mud on their faces . . . steel-hats pulled down low on their foreheads . . . and they were talking in subdued and earnest tones . . . Rather condescending, I thought . . . No arbitrary commands . . . down there in that dug-out, facing . . . they and we . . . knew not what! All men had reached a common level before impending death . . . And I was happy I had seen these men there, but . . . (The captain speaking again) . . . "Now, men, there is little to be said to you. I understand, however, . . . This is correct is it not, lieutenant? . . . our company has not drawn its quota of hand-grenades for your squads . . . and that will be all, men."

We started filing out. No salutes . . . just men leaving, and officers sitting on the muddy floor. I reached the blanket over the door . . . "Corporal Emmett, just a moment. Get some man to get your grenades. You have not had any sleep . . . Think you'd better sleep a little." . . . "Sir, with the captain's permission, I prefer personally to see to it that my men receive their grenades . . . I consider it rather important . . . You see, they might need them;" . . . "Damn fine spirit, corporal . . . but get all the sleep you can . . . and 'Good Luck'!"

Almost staggering out into the darkness, and now blinded even by the weakness of the candle light, the night's darkness was impenetrable to me. I held out my hands and walked straight ahead. I could not find the sides of the trenches. I could see nothing. Perhaps I took a dozen steps before I waked up to a realization that I had gone past the turn to the left. I wheeled and felt for the trench-wall, intending to retrace my steps to the mouth of the dug-out and try again. But I could find nothing familiar to me. In confusion, I walked again. Now I could not find the three intersecting trenches . . . I stopped a moment to think . . . and it was peculiar how still everything became to me! No sound by which I could direct my course came. But I was confident I knew the direction, and I walked again . . . Not finding any 'land-marks' I did a foolish thing. I crawled out of the trench, and scrambled along through the briars. They cut my face. I fell down. Again to my feet, and I fell. Then I knew I had gone to sleep while walking . . . and was lost! I sat down . . . listened, waited. The big guns groaned far away; then they sprayed the woods with shrapnel. I realized I would be hit if I sat there but tried to keep from getting panicky. I heard a noise and went toward it. I called softly. An answer: "Emmett, you infernal damn blockheaded fool! Whatcha doin' up there. I started to shoot your fool head off. Think I'll do it anyway. Get down here before someone else kills you . . . This way, you chump. Are you lost?" . . . It was old Charlie King. I jumped, hitting the bottom in the mud, and King gave me a resounding kick . . . "Wake up . . . Now this way . . . Straight ahead, and you'll find your men there." . . . and putting a friendly arm around me, he patted me on the back, . . . and I found my men.

All the men had the news. Packs were being rolled. Many questions were being asked . . . Excitement everywhere . . . Now I remembered we had no hand-grenades, and off I went to the 'dump' in the darkness, arriving more through intuition than direction. An officer was just finishing issuing grenades. I took a supply and went back to the boys who eagerly took them now. Back in the days of training the men would handle these bombs gingerly. Now they snapped them onto their belts with a flourish of satisfaction.

But I was 'all in' and I started toward my hole in the wall and told Clarence to "wake me up when the chickens begin to crow." . . . It is peculiar how men will continue to jest even when facing the uncertainties of life! . . . "Before the chickens begin to crow", repeated Clarence. "Yes, they'll crow loud and lay eggs tonight" . . . I pushed into the side-wall and sank into unconsciousness . . . Someone tugged at me. It was Clarence. "Well, old top, better come alive . . . Chickens are laying eggs now, hear 'em? . . . You've slept thirty minutes, and Jerry is desirous of making your personal acquaintance."

I staggered up and found the men moving in the trenches. Clarence helped me slip on my pack . . . "Why will they make us fight with these things swung to us?" . . . We made it into line, and holding to the gun ahead, plodded along toward the north . . . into the woods . . . down the hill. O, how dark it was! . . . I recognized the hillside of the previous evening-exploration. I came to the overhead covering of the trench . . . Man behind man, each man holding to the gun of the man ahead . . . slip, slide,—and there was no sound except the sludge . . . sludge . . . of muddy feet.

Had anyone on the night of September 12, 1918, suggested to me that some day in connection with the events transpiring I would quote from Wilhelm, Crown Prince of Germany, in explanation, the idea would have been too preposterous for comment. Now, however, time has intervened. And what changes time makes![1]

(Wilhelm, Crown Prince of Germany) :—

"Brave men everywhere recognize heroism, even in the ranks of the enemy. Even now there appears a heroic fellowship which unites brave fighters on every front and in every land. This fellowship may prove more powerful than political and commercial divisions and alterations. It teaches mutual respect and mutual understanding. I, for one, expect more for international reconciliation from the men who have gone through the inferno of war than from the men who, seated behind mahogany desks, direct or misdirect the fate of nations.

1 (Ed.) The first day of the St. Mihiel Offensive to reduce the German salient that jutted deeply into France towards Paris. This was the first major offensive of the American Expeditionary Forces and also their first significant victory.

"Having known the stark reality of war, I can visualize its horrors as well as its sublimities. No pen, no brush has ever depicted a battle adequately. No book, no picture, that I have seen, conveys the terror and grandeur of the battlefield . . . But somehow the awesome spirit that pervades the battlefields has escaped . . . Neither . . . can suggest even remotely the awesome figure of Him who faces us on fields of carnage and spreads the mantle of His majesty over the multitudes which He blesses even as He robs them of life . . . The experience of the battle field somehow brings man face to face with his destiny and his Maker . . . It is something that every soldier who has faced death has felt." . . .

Whatever is that "awesome spirit" its presence was with me as I went down into that valley . . . a valley of anguish for all . . . a valley of excruciating misery for many . . . a valley of death for some.

We passed from under the canopy of the trench near the bottom of the hill. The impenetrable darkness forced tired minds to a consciousness of impending doom. Faltering in step, and fully aware of the importance of the moments, I shook off the spell and looked about. Not even the man next to me onto whose gun I was holding could be seen. I marveled at the completeness with which we had been enveloped by the night. As I stood still, straining to peer through the murk with face turned to the southeast, almost above me on the hillside, a crimson flame, like a long arm, struck out into the darkness. "Pink" came the almost timid sound to our ears. It was the signal-gun for the barrage. The flash had little more than been absorbed by the inky darkness when a crimson circle sped along the entire front to our rear. Crimson was followed by the roar of belching guns. At first we could hear the shrieking shells as they passed high over-head Germany bound. But the second volley followed so closely upon the first discharge that the sound echoed back upon us, to be repeated over and over again until we were fairly rocked with the reverberations. We stood in an inferno, our ears no longer registering sound, our eyes no longer capable of encompassing the intensity and luridness of the light. A grander sight no man ever saw! No more stupendous spectacle ever smote the eyes and ears of man.

Intoxicated with its awesomeness, startled with its suddenness, I thought to check the accuracy of my observations. (Why will men do such strange things under stress?) . . . And I took from my pocket a bit of newspaper. I actually read it with ease by the flare of the guns. By then the line of fire had skirted the entire skyline to our rear in both directions as far as the eye could reach. A solid red stream of flame flowed toward Germany!

Momentarily noting the complete absence of shells coming toward us from the German lines I concluded our own barrage had been so devastating and complete that all enemy guns had been silenced. A feeling of relief was pervading me . . . but the duration was short.[2]

Slashing back at us came the boom of the German guns. Boom for boom, shot for shot, and the earth reeled under our slogging feet. Derivation of sound was no longer determinable. Speeding shells shrieked, cried, and whined above us. The whistle of flying steel split the air in its flight. Lurid flashes brightened up an already crimson sky, and we . . . deep in our recess of comparative security in the valley below the hill . . . looked upon the cataclysm, not as a part, but as detached observers unable to realize its significance.

Troops ahead of us, having blocked our progress by their slower ascent of the hill, now moved out more rapidly and we proceeded. Ross was in the squad just ahead of me. I do not know why,—I do not suppose he does either,—but we quarreled again. I could not see him but, reaching, was fortunate enough to catch him by the throat . . . I choked him. I cursed him, depleting a fervent vocabulary . . . Then someone threw us apart . . . We went onward.

At the crest of the hill a machine-gun company intersected our company, mixing up the men. The 'boys from the death-squads' did not

2 (Ed.) "Barrage" has several meaning to the troops in World War I. Because of the field telephone and wrist-watches, war now operated on a matrix. Friendly artillery would all fire a mass barrage at pre-planned targets at the same time across a sector to "soften up" the enemy defenses at the front lines or the supporting artillery beyond. Often an assault would be supported by a rolling barrage of a line of shells fired across a trench and lifted to another linear sheaf a hundred meters of so beyond on a pre-determined schedule. Sometimes, the artillery would fire a creeping barrage where the guns added fifty meters or so with each shot and the attacking troops followed just out of shell fragment range to cross the battlefield before the machine gunners rose out of their bunkers and resumed firing. It was a risk for advancing Infantry to follow the barrage too closely, but it was more likely to be cut down by enemy machine guns and rifles.

seem to know where they were going, but each gunner seemed to be vying with the others to determine to whom should go the trophy for the most adroit use of profanity . . . But it was all impersonal, directed against the impediments, chiefly against the heavy machine-guns.

Disentangling ourselves from the mix-up we kept up the sporadic march. The trip of the previous evening now came back to us. I could see its importance. The flare of the guns, which had now settled down to a steady, mechanical grind, no longer furnished light for travel. Darkness was upon us, except for the occasional flash nearby which lighted up our surroundings with a pinkish hue, distorting the shape of all things before us . . . And I was now thankful I had been on the 'evening party' for I could lead my men along the trenches in the right direction. We were now far out into No Man's Land, and I turned to the left and spread out with the other men in a deep trench, and, along with the others, fell exhausted into the mire.

The slackening of the fire gave us hope there would be a surcease of the gruelling misery . . . but now it began to rain. The big guns again took up each other's challenge, and with a meticulous regularity shot for shot flew over our heads. The crack of each gun seemed to drive down upon us an increasing down-pour. The quiver of the elements following the explosions shook the water from the skies like drops from the leaves of trees. Water began to accumulate in our trench. We arose to our feet leaning against the muddy banks. Our hands were slimy with the oozy clay. Our feet now felt the chill of the penetrating cold. I sat down in the slush, again, no longer being able to hold the weight of my body on my feet. The mud was cold and the water seeped through my dirt-encrusted uniform. Shivers ran over my body . . . and I wanted to do something . . . go somewhere . . . just anything . . . anywhere!

The cannonade had now settled down to a steady pound, pound. I lay under the trajectory of fire and, when on occasions I peered over the parapet, could see the flash and bursting of shells both to the front and to the rear, the German fire upon our batteries. The hours of the night wore on. With each minute the piercing sting of cold made me more and more uncomfortable. My feet began to feel numb. The rain upon my face no longer inflicted a punishment. I found I was huddled upon the ground, partially asleep, my head being supported by my left arm, now mired in the mud with my face covered by my 'tin-lid.' Nature in her march had rendered me immune to further pain. Ross crawled over me. A recollection, at the sound of his voice, was revived. I had treated him scandalously . . . and for that I was sorry . . . I yelled my regrets into his ear, pitching my voice above the booming of cannon . . . and he laughed . . . "What the hell you talking about? . . . If you did anything to me I don't know it." . . . This was likely the truth . . . A little friendly fight, under the stress of conditions, had not registered on his distraught mind!

I lay down again on the ground only to be conscious of the fact that water was covering about one-half my prone body. It was a wretched place.

There was a sound of the movement of men. A faint light began to appear in the eastern sky. An officer approached. "Corporal Emmett" . . . "Here, Sir." . . . "You will take charge of a half-platoon of men . . . Place someone in charge of your squad." . . . "Clarence, old boy, will you follow me?" . . . "Follow you? . . . Why, I'll follow anywhere you'll go!" . . . "All right, Lieutenant. Private Clarence Oliver in charge." . . . "Now boys, remember . . . Lead out at five . . . Exactly . . . Keep spread out . . . and Good Luck."

Someone was tugging at my arm. The bulky figure of old Surrat now was partially visible. "Corporal, have we time? . . . I have to get out of here a minute before we go . . . O, hell! Too late now" . . . and the tension was relieved as the whole squad laughed at his embarrassment.

The rain became lighter. The sky now reflected its faint glow toward the ground. The movement of people could be seen, and, as if set off by one electric fuse, there was a spontaneous crash of cannon to our rear . . . The morning barrage had begun!

Shots were falling just a few yards ahead of our location. Dirt flew high in the air. Puffs of flying mud overlapped. Bursting shells ripped the air. Shrapnel whined far out over the maze of trenches. I ventured to peer over the trench-top, and thought for the moment that the barrage was solid so close together did they fall. Now the barrage raised and the explosions rent the earth some two hundred feet farther away from us . . . That territory had been 'blasted out' . . . I looked at my watch. The hands pointed "5" . . . *The Battle of Saint Mihiel zixis to begin!*

Men were rising from the mud all along the trenches. I looked to the left . . . A soldier was moving from his concealment into the open. He stood erect, limned against a faint sky-line, his rifle grasped in his left hand, and, without taking a step, the rifle fell . . . Slowly he bent over . . . just perceptibly . . . Down on his face with a thud! He was dead . . . dead from a machine-gun bullet. Momentarily I could not understand what had taken place so suddenly before my eyes.

I stood with my head above the parapet. Old 'Lord' George was crawling up the bank. I caught his foot and 'boosted' him, and he lay flat on his belly. Reaching back for my extended hand, he yanked at me. "Zip . . . zip . . . zip . . . pst . . . pst" . . . a peculiar noise, like a fast flying insect, almost touched my face. I struck at the 'insect'. "What the hell, 'Lord', is that biting at us?" . . . "You damn numbskull. Get down or you'll never find out . . . Machine-gun bullets."

I planted myself on top of the parapet alongside 'Lord' George, and lying flat for protection against the zipping bullets, now sizzing all along the line, caught the extended hands of men. We dragged them up. Then, "Spread out, boys . . . Let's go."

We crawled along a few yards while the pst . . . pst . . . pst . . . of steel-jackets flitted a few inches above our backs. Just ahead I saw a barbed-wire entanglement, a bramble impenetrable. I arose to my knees. From my belt I slipped a wire-clipper and with a whack . . . whack . . . whack . . . I worked my way into the tangle. Soldiers were crowding all around me. The entanglement had 'bunched them up'. We were penned in a wire-ambush, and the 'zing' . . . 'zing' of machine-gun bullets, glancing from against the wires, showed us the German knew our predicament. It was a perfect trap into which we had crawled unsuspectingly. I tried to move out, and tripping on a wire which snared my leggings, fell to the ground, slashing my face as I went down. Blood trickled down my cheek . . . "O.K. . . . Nothing serious".[3]

3 (Ed.) "Steel jackets" refers to the thin steel coating over the lead core of the rifle and machine gun bullets.

A machine gunner burst a volley of steel around us, and I began crawling back over the course of my advance. Rising to my feet I ducked involuntarily,—luck being with me for I fell into a large shell-hole, as a big mortar-shot, wobbling with its energy almost spent, cleared our heads only by feet. Instinctively I turned to watch its progress. It thudded into the soft earth a few yards to the rear. From it oozed a thin, green smoke. Just beyond the fallen shell, and coming toward the salubrious gas, was Clarence. He was running toward me, unmasked . . . The characteristic grin was on his face. Suddenly he stopped . . . Wavering a moment . . . weaving . . . he went to his knees . . . over on his face . . .

Lieutenant Nysewander, now with us in the protecting hole, was shouting . . . shouting . . . shouting sharp shrill commands. Around him had gathered a group of his men . . . until now, more a mob than a disciplined company of soldiers . . . But we were being held up by the wire . . . "The wire . . . out there . . . Who has a pair of snips . . . Cut the wire!" . . . Corporal Jesse M. Grisham stood up. I would have warned him, for I had been out there cutting that wire, but he looked at no one. He was walking erect, leisurely, ahead. He knelt at the bramble. We could hear his snips as they ate into the trap . . . Then . . . he relaxed, and folded over . . . The War Department, now, has the record:

DISTINGUISHED SERVICE CROSS

GRISHAM, JESSE M. . . . Corporal.—Deceased.

Company L, 359 Infantry. For extra-ordinary heroism in action near Fey-en-Haye, France, September 12, 1918. When the advance of his company was halted by an impassable barbed-wire entanglement, he voluntarily jumped out of a trench in the face of heavy machine gun fire and cut sufficient paths through the wire to enable the company to continue its advance. In the performance of this self-sacrificing act this gallant soldier was killed.

Next of kin, Mrs. Mary hockey, mother, Holland, Arkansas.

Lieutenant Nysewander, now, was tying a handkerchief . . . a white one . . . to a walking stick . . . Could I believe my eyes? . . . "Follow me . . . You can tell by this." . . . He held it aloft. What a spectacle! And what a leader. God have mercy on his poor soul! Our line straightened, momentarily, under the spell of his action, and we surged forward. More wire-fences . . . a veritable bramble . . . rolls of wire, spikes, wires stretched tight, wires in every shape, stood to check us! Hardly had the humor of the lieutenant's 'flag-of-truce', 'here-I-am-shoot-me' foolishness flitted from my mind until it looked serious. I felt for my clippers. They were gone.

Unable to extricate myself by cutting through and knowing that a pause courted death, I decided to go straight through . . . and over . . . regardless. Placing a foot as high on the strands as I could step, I lunged and jumped. A wire caught my left leg about even with the hip. I struck the wire-covered ground on the other side of the entanglement . . . minus a large and important section of my pants. Wading in wire and being held back by our soldier's equipment, I slipped my bayonet and cut loose from everything except my 'iron-rations', belt with shells, and rifle. Now free from some of the burden, I found I could move through the maze with some semblance of ease. I worked forward through a clearer space. I ran rapidly toward the right, circling the worst of a wire-barricade, and came out into the open-ground . . . but not without having passed a number of khaki-clad youths who were draped over the cruel, gripping wire . . . Blood ran . . . drop . . . drop . . . drop . . . crimson splotches upon a sodden field of carnage.

Now I saw that by my deflection I had become separated from the men of my company. Far to the left . . . up and down . . . in and out . . . was Lieutenant Nysewander's flag!

Our progress had been faster than the orders-of-advance had contemplated. We were now almost against the falling shells of our own barrage. Shrieking shells burst above us. Shrapnel slashed back at us. Dirt spurted high in the air and settled down upon us. We were stifled by the acrid pungency of the vapors from the explosives. Some-one yelled: "Our own barrage! God help us; we are in it! . . . Who has a Verey pistol? . . . Damn it, shoot a signal . . . Raise the barrage." A nearby officer, unperturbed, arose to scan the effect of the falling shells. The signal went up, and soon the crash of explosions was beyond and out of our range.

I had gained a position at the foot of a slightly rising elevation where I was comparatively free from machine-gun fire. To the left, particularly, shells were bursting in the greatest profusion. In that vicinity the wire was the densest, and our gunners were concentrating their fire to blast out the strong-hold. There I saw a most interesting spectacle. Our guns were throwing 'fish-tails'.

A 'fish-tail' is a shell to which is attached a staff. The staff has affixed to it a number of barbs, or hooks, which, when shot on the level with the ground, are intended to catch into the wire entanglements and drag down the barricades. The speed of these fish-tails did not appear to me to be very rapid. They bumped along on the ground, striking, rising, striking, . . . and sometimes ensnaring a wire but evidently serving no particularly practical military purpose. I concluded they must have been born out of the fervency of mind of some distinguished ballistic expert who had received his inspiration from munition dispensers.

I scampered from hole to hole, sometimes running, sometimes crawling, taking advantage of 'cover' wherever possible. In one of my leaps into a hole for safety I fell upon my friendly-enemy, Corporal Ross. Upon my arrival, he raised up, then fell backwards, whimpering: "I am shot". His pallid face convinced me he was mortally hit. I stripped his pack for him, searching for the tell-tale blood but found none. "Where are you hit?" . . . "Here," and he indicated his left shoulder with his right hand. I ripped off his jacket, and saw nothing. Then I stripped down his shirt and a small trickle of blood oozed from just below the clavicle. Bending him over so I could see the point of exit of the bullet, I found a clean wound bleeding very little. Feeling over the bones, I found no fractures, and as he bled little I concluded he was not severely wounded. He contended he was going to die. Consoling him the best I could while I took his emergency kit to stanch the trickle of blood, I left him lying flat, scared, and pale.

Typical wire-entanglement immediately to right of Fey-en-Haye throughout which the 90th attacked.

A battery now turned its attention to the immediate vicinity of my discomfort, and I shot head-first into a convenient shell-hole, landing directly on top of old Bill Thompson. He was sitting spread out in the mud, laughing at the top of his voice: "Hi, there, Emmett. This is the most fun I ever had."

The minds of men react differently under the same circumstances. Some men, when under fire, cringe and dread to expose themselves. Others become obsessed with the tremendousness of the thing and stand mute; still others are fired by the furore and take an hysterical delight in the adventure. I concluded Thompson belonged to the latter class.

The two of us sat for a few moments until the battery shifted the fire and we became inquisitive, showing our heads which resulted in bringing the fire of a machine-gun immediately upon our position. "Chat . . . chat . . . chat . . ." and the dirt flew over us. His aim was directly on the rim of the hole, a very uncomfortable position for us. Two more men found our refuge and our hole was full. Another industrious German then concluded to impress us with his importance and added fire from another angle. The edge of the hole was rapidly being trimmed down. Our faces were full of dirt, and the prospects were un-alluring!

With the hole full to the very brim with our four wriggling bodies, each adjusting himself with every opportunity 'toward the bottom', Bill suggested that some of us "are going to have to leave here . . . and I don't mean 'tomorrow' either!" . . . "Well", I said, "we preempted this particular sector, but I don't care much about it,—especially just now; and judging from what I can hear, I don't think this war's quite over, so let's get to hell out of here and engage in a little more of it." That was the decision. Our compatriots did not seem to be disappointed when we left them alone.

It is rather uncanny how one learns the mannerisms of an enemy and how we learned the German's use of the machine-gun. We made this observation: Cartridges for the guns came in 'clips'. Each clip held five shells. These cartridge-containers are slipped onto a belt, the belt being fed into the gun. When a clip is fed through there comes an almost imperceptible break in the rapidity of the fire as the gun takes up the following clip to begin firing it. This 'break' seemed to be a signal to the gunners to pause. This resulted in the German firing a clip and automatically pausing, presumably to note the effect of the volley. This gave the peculiar "tat . . . tat . . . tat . . . tat . . . tat", then after a pause, the sound would chatter back to us again.

Wishing to leave the hole, Bill and I, instinctively, waited for the pause. It came . . . and Bill ran. Not expecting such a sudden appearance of an American soldier, Bill was enabled to fall into a hole yards away before the familiar chatter came again. And when that burst had finished I surprised him again by a repetition of the performance. I was not quite so lucky, though, for dirt was flying all around my feet before I slid 'Kelly-wise', safe and sound but scared half to death, into another of the innumerable holes.

Spreading out like a fan, I thought to rest a moment, but a big 'G. I. can' broke above my head, raining chunks of hot iron all about me. My left hand was lying flat on the mud. It stung . . . and I looked. From my hand at the joint of the index finger a trickle of blood was showing. Between my finger and thumb was a hole in the ground into which I probed. I extracted a hot piece of steel about the size of a small marble. I licked the blood from my hand and, realizing I was not badly hurt, shoved the steel into my pocket as soon as it cooled sufficiently for me to handle it . . . and "Well, that's a little souvenir, anyway."

Another gunner picked up my position and forced me low into the hole. His bursts of fire were rhythmic, a burst of five and a pause,—over and over again,—and the dirt dabbed me in the face. Seeking to protect my head I cupped my helmed over it and lay still . . . an interminable time, it seemed to me . . . In reality it must have been only a few seconds. Finally the gunner stopped short. I concluded he thought he had killed me . . . And then a peculiar coincident happened!

Just to the fore of the hole in which I concealed myself was a triangular piece of steel, a steel-stake which had been used by the enemy onto which to fasten the strands of wire of their entanglements. The 'V' of the angle-iron had its point away from me, or the sharp point was toward the gunner. Thinking the German had deserted me, but being unwilling to place an implicit trust in him,—in that year of our national tranquil relations,—I decided to investigate,—but with caution. I raised my head slowly and carefully above the level of the ground . . . "Pop" . . . and a bullet struck the iron-post within six inches of my face and on a line with my eye. The flying rust sprinkled into my eyes while the bullet, deflected by the contact with the post, 'zizzed' past my ear. I saw the mark on the post . . . and fell back as if dead . . . Long did I wait before I raised my head again. Evidently the German saw he was unable to reach me in the hole with the machine-gun and decided to catch me with a rifle-bullet when I appeared. And seeing me disappear so quickly after the rifle-shot he must have concluded I was an addition to the "recent departed", for when I did decide to come out of there I proceeded some considerable distance before being shot at again.

Now I ran to another hole and crawled around to a point where I could see a wide stretch of open country being swept by machine-guns. It was a very uncomfortable location, indeed. But I looked it over and decided if I could reach the trenches at the top of the incline without getting shot, then, I could go 'anywhere', and I started on a long run. A veritable swarm of 'insects' zipped after me. At last, I fell into a trench . . . safe!

I was no longer alone. Many men had taken refuge in this depression. I related the manner in which Bill and I had outwitted the gunner, and four of us decided to repeat the performance. Out went a lad . . . and he made the trench ahead. I went next. I reached the trench, and to my horror found it filled with up-turned spiked-shafts. To have jumped onto one would have meant to have been impaled there. But my choice was to jump or be shot. I jumped, landing against the farthermost side of the trench and missing the points only by inches. Looking up I saw another boy standing, hesitating to make the choice. I called to him: "Jump, damn it, jump!" . . . "Spat . . . spat . . . spat," . . . the sound came to me, and he crumpled and fell. I caught him, shifting his weight from the cruel barbs, and an unknown partner helped me to place him on the ground. A large splotch of blood appeared in his groin. We cut away his pants, and there was a bloody hole into which I might almost have placed my fist. The blood was spurting. The wound was so close to the hip that little could be done to stanch the flow of blood, but we ripped away his emergency-kit and padded the wound . . . and then the rain began to fall again . . . More men were now with us, and we dug a shelter for him into the side of the wall . . . The pallor of death was showing in his face . . . He was shivering, but we could do nothing but place him on the ground . . . Another soldier came along wearing a raincoat . . . I asked him to let me have it to cover the dying man . . . He said he needed it himself! . . . What a soldier! What an American! . . . Slipping my bayonet from my gun, I jabbed him as he fled. Another boy having heard and seen aided his departure by a continuation of the jabbing . . . jab . . . jab . . . jab . . . until he was forced over the top of the trench into the bullet-swept field . . . Knowing the injured buddy was beyond assistance, such as we had, I leaned over him and wrote his whispered name on my shirt-cuff; then walked away . . .

Down the trench a few yards I encountered a stern-faced American lad who stopped each approaching man and forced him on bended knees to crawl beyond a trench-entrance which opened toward the enemy. A hard-boiled 'brass-hat' came upon this lad, swearing: "You damn coward . . . Get away from there . . . Let us pass" . . . Without emotion the soldier replied, "Lieutenant, I am no coward. To go into the mouth of that trench means death," and turning back his raincoat we saw his blood-soaked jacket. His chest was literally sieved by bullet-holes. And as he showed the results of his having protected us, he swayed . . . and fell dead.[4]

We had just had the exemplification of two sorts of men, the same kind of men who fight wars and live and die, those who compete in the workaday world and are loved and hated, who give and take. After the turmoil of battle had passed and my mind had grasped the import of the acts of those men that day, acts which showed their characters, I could not but think of them in terms of the Texas poet who said:

"Heroes.
"One dared to die. In a swift moment's space
Fell in War's forefront, laughter on his face.
Bronze tells his fame in many a market-place.
Another dared to live, the. long years through
Felt his slow heart's blood ooze, like crimson dew,
For duty's sake, and smiled. And no one knew."

The men of the 90th Division had gone over the top in two 'waves', one long line followed by another after an interval of space. Encountering heavy fighting ahead, the first wave was slowed down until the two lines of soldiers merged.[5]

With the counter-bombardment all Americans took to cover in the trenches like veterans. An enemy aviator sped over the field and located a

4 (Ed.) The term, "Brass hat," was Soldier slang for a staff officer. It was sometimes was used to denote any officer that the Soldier had not met.

5 (Ed.) For hundreds of years Infantry units advancing in close order or skirmish lines have advanced in successive waves or separated lines. This is to keep the units following from getting intermingled with the lead wave or to allow bullets or shells that missed the first wave to pass over and fall in the space between units and not strike the following units. In the days of musket balls with a rainbow trajectory, this cut down on battlefield casualties. By World War I, this minimized the casualties caused by one bursting shell, or machine gun burst, caused by troops bunching up..

long trench filled with infantrymen. He saw his chance to hurl destruction into our midst, and banking for position, slowed his plane and pointed its nose along the trench. His machine-gun was spitting at full speed. Momentarily the men in their consternation at the turn of affairs sought cover against the side-walls. Then, as if acting with a single thought and common purpose hundreds of infantrymen arose along the trench and began pumping volley after volley toward the daring airman. Great pieces of the plane broke away and floated into the air. Then the German wobbled slightly, and spinning tail over nose the maltese-cross fell carrying the plucky enemy to his death, burying itself deep in the mud. Perhaps far back of the Rhine some mother was pleading:

"Drink of this cup—when Isis led
Her boy, of old to the beaming sky,
She mingled a draught divine, and said—
'Drink of this cup, thou'lt never die!'
Thus do I say and sing to thee,
Heir of that boundless heav'n on high,
Though frail, and fall'n, and lost thou be,
'Drink of this cup, thou'lt never die!'

Somewhere in the fight that morning we came upon an officer who had rallied around him a sizeable squad of men. With great skill and caution he was directing their movements. In our efforts to progress we were running afoul of one machine-gun nest after another. The enemy had placed these gunners in concrete 'pill-boxes' which were impregnable except for heavy cannon or grenades thrown into the sunken rooms through the doorways. Inasmuch as the Germans were firing their machine-guns through narrow slots just above ground, it was very difficult to approach a pill-box close enough to cast a grenade through the door. With coolness and deliberation this officer took upon himself the task of eradicating these pill-boxes, placing us here and there under cover with instructions to surround the box completely and make the German continue his fire by harassing him with our rifles. By doing this one of us would get close enough to 'blast him out' with a grenade.

I lost contact with this officer in one of these sallies. More than an hour later I came upon him lying part-way in a trench. Stamped upon his face was the blanch of death. Uncomplainingly, he told me he had been shot. Two small holes were in his canteen yet hanging from the hip. The water was trickling through the holes, showing he had just received his wounds. I pulled him to a position of comparative safety where he stretched upon the ground. Then pandemonium in its most violent form was ushered upon us. Battery after battery of 'whiz-bangs' churned away. Shell after shell broke above us,—just in front of us,—and we would have been wiped out had we not taken shelter against a bank some fifteen feet high. So direct was the fire that, with each exploding shell, tons of dirt shifted into the hole threatening to bury us alive. There was nothing we could do for the lieutenant under the conditions and every man began to care for himself. Realizing the precariousness of the position we had selected, I went westward out of the shelter to a point where there was little depression. A machine-gunner 'picked me up' and I went flat down and crawled . . . and what a sight was before me!

A large shell had made a direct hit upon four boys. All were dead. Limbs were mangled . . . bodies were torn. It was a sight revolting and beyond description. Of one of my comrades I could find only small fragments of his poor body . . . None were larger than my hand . . . with the exception . . . there lay his head, jerked completely from his body. The skin from his neck was stripped back to the crown of his skull . . . Bare white bones of his head, smeared over with a pinkish thin blood not yet congealed glistened in the light. The powder-blackened face of a young Jewish boy stared immobile into eternity. Nearby was his hand which had been popped off the arm just back of the wrist . . . On the bleeding stub was a wrist-watch . . . And I looked at the others . . . Their spent, distorted bodies, with muscles yet twitching, were mute condemnations of man's civilization! And as I crawled back toward the wounded lieutenant over the mushy, disintegrated, fragmentary parts of four of my buddies, with their blood glutinous on my hands and knees, I prayed God for the time to come when the men who fight our battles would banish our commercial-lechers or cause them to be the first to fight.

I found the wounded lieutenant writhing in pain. A sergeant came along. He was rather old for a soldier in the ranks. He was florid of face, and his neatly trimmed red moustache showed in contrast to his fast greying hair. I called to him to assist me with the lieutenant. Indiscreetly he stood up. With a grunt he sat down, gurgling: "I am shot . . . Through the chest." A gas-shell popped above us and we stuck our heads into our masks. Looking at the freshly wounded sergeant, who was now sitting with his head in his hands, I realized he was unable to put on his mask and would surely die without it. I grabbed it. To my horror I found the pipe leading to the inhalation-tank punctured by the bullet which had entered his chest. Frantically I repaired the pipe and fitted it to his face. When last I saw him he was leaning over, breathing laboriously. A frothy, pinkish spume blubbered from his nose and mouth.

Not content to be cooped up in that pit I worked northward 'toward Germany,'—despite the impelling urge to take the opposite direction. I had gone but a short distance when an obliging sniper potted a few shots at me and sent me scurrying to a dug-out. I would have entered had not a wire struck my foot. Instantly I halted to explore. The wire ran across the bottom of the doorway just high enough from the lower step for one to trip his toe under it. From across the step it passed along the casement of the door to a point even with my head . . . just a few inches away . . . There it was: a hand-grenade fastened to the door-casement. It would have exploded had I tripped the wire . . . I retreated toward the point where we had had comparative security, and as I went I muttered violent, malefic curses upon a cunning enemy who would attempt to destroy life by cunningness, and all the time I was thinking of some ingenious way to secure revenge . . . "Vengeance in my heart, death in my hand."

The fury of the battle had now swept farther ahead. Shell-fire was less severe generally over the field, and I had time to notice the pangs of hunger. I slipped out of the depression and scanned the country to the northwest. There I saw a long line of American soldiers entering a skirt of timber. They impressed me with the deliberation of their movements. I could not suppress the thought that they looked like hunters beating out wild beasts from their concealment in the timber.

Merk now joined me. At first he expressed a great joy in seeing me. He had heard, so he said, I had been killed. Then in his genial, good-natured drawl he added: "To bad; yes, too bad. I've been telling some of the boys . . . and they'll think I lied!"

Now we sat out in the open where a better view of the storm-swept country could be seen. Our lips were parched from piquant battle-fumes, and we trembled from exhaustion. Here we were joined by a whimsical wag who was all bespattered with mud. From the corner of his mouth drooled a goodly portion of Uncle Sam's last issue of chewing tobacco. In no uncertain terms he expressed a desire to eat; then peering ahead of him at the inert form of one of the boys 'gone west' on whose back was strapped his 'iron rations', our cynical friend began quoting from Milton:

"Straight mine eye hath caught new pleasures" . . .

Retrieving the rations from the back of the dead buddy with the comment that a 'chinaman's funeral won't help him any', Merk and I assisted him to devour the food.[6]

A major appeared on top of a pile of stone to our rear. His face was familiar and I went to him. I recognized him as Weathered, a Texas University class-mate. Neither of us spoke of our past acquaintance. We talked of the fight, the reorganization of the men, and who had been killed. He wanted to know where were the wounded, and ordered a dug-out to be cleaned and the wounded brought there. Two stretcher-bearers came along. I told of the boy with the big hole in his groin . . . They went that way, but the litter came back empty. They said he was dead. I spoke of my lieutenant friend with the shots through the hips . . . And they went with me . . . Someone suggested he might live if he could get medical attention. And

6 (Ed.) In a "Chinaman's funeral," food and other items would be buried with the dead to comfort them in the afterlife.

when he heard these words a great fright came over his face. A suppliant tremor came into his voice: "Take me to the doctor, boys, please." . . . "O. K., Lieutenant: Let's go," and we laid him on the stretcher . . . Tramp, tramp, tramp, we went with him toward the rear . . . Gunners opened up on us again, now that we were showing ourselves, and we retreated to a trench for protection and concealment. It was hard walking carrying the stretcher down there. We were constantly coming to rolls of wire through which we could not cross. We went out into the open again.

It is peculiar what effect pain will have on the actions of man! Here was a soldier, fearless and undaunted, until struck by bullets. Writhing now with the excruciating pangs of his wounds, he feared the exposure would cause him to be wounded again. He became panicky, and grabbing his pistol from its holster flourished it at us, commanding: "Get back into that trench, damn you. Don't take me out again. I'll shoot any man who tries it!" We laid him on the ground and tried to cajole him into an understanding of the necessity for our movements. Our words made little impression. We told him we were his friends: that only his gallantry was causing us to return him to the 'medicos'; that we would not disarm him, for we respected his position as a brave officer. As I looked into his once fearless eyes, now cowering and craven under the scourge of pain, he relaxed his grasp upon his pistol, and I wrenched it from him. He was now within our power.

Whilst we wrestled with the wounded officer an enthralling matching of wits was going on a short distance beyond. In our company was a long, gangling, freckled-faced, serious-minded chap of the Alvin York variety. He, too, was adept with the use of the rifle. I saw him moving cautiously along a protecting bank. He would peer, move, and look again. He took his steel-helmet and placed it on top of a bayonet, just showing it above the ground-level. "Tink" came the sound of a rifle-bullet from a sniper's gun as the bullet struck the decoy-hat. 'Old Freckles' noted the course of the bullet as it ricochetted beyond the 'derby'. Then he moved along several yards and put up another hat. Again the sound clinked off the decoy, and 'Freckles' checked the course of the bullet. Now he moved around to a point of vantage where he looked long and carefully. Suddenly he arose from under 'cover' and fired. From out the top of a tree, some two hundred yards away, fell the form of a German. Freckles had located his quarry![7]

We pushed along with the lieutenant occasionally meeting other men, some carrying stretchers upon which lay inert forms, and some dragging their blood-spattered bodies the best they could. Our progress was slow for our bodies were weary. Every step, every little jolt, sent the pangs of the damned through the shattered body of our friend. Carefully, painfully, we moved with him. Two more slightly wounded men came to us. They were going back with a 'blighty'. They volunteered to help . . . And six men, now, exhausted almost beyond further physical exertion, relaying each other yard by yard, went on with the burden. Machine-gun fire no longer molested us. We were too far to the rear for that, but the field was being raked with high explosive shells. There was nothing we could do about them, and we got out into the open . . . unconcerned . . . "O, what was the difference if they did hit us!"

The terrain dipped downward. We saw the valley below, that valley through which we had passed during the night just gone, that night which now seemed years and years ago, when amidst the scream of slashing steel we had paused long enough to turn our eyes from jets of flame to look heaven-ward and say:

"From this hour the pledge is given,
From this hour my soul is Thine:
Coma what will from earth or heaven,
Weal or woe, Thy fate be mine!"

A crowd of men hugged close against the bluff. Carefully we bore our burden down the embankment . . . "An officer, Captain", I said nodding toward our litter. "Pretty bad hit . . . Through the hips." . . . The captain opened the wounded man's eyes with his finger-tips, and with a "You'll

7 (Ed.) Medal of Honor awardee Corporal Alvin Collum York, Company G, 328th Infantry, 82nd Division. One of the most decorated, if not the most decorated American Soldier, in World War I. He was a famous man inside and outside the Army even before the award-winning film, "Sergeant York," premiered in 1941.

make it all right, old fellow", said to us: "Take him over there with the others . . . Give him a 'shot' sergeant!"

We turned to leave but a feeble voice quavering with emotion called after us: "Thanks, boys. You're a damn fine lot" . . . and tears came into his eyes and ran down the cheeks of a powder-smudged, anguish-wreaked face. . . . A resounding blow struck me in the back . . . "Well, I'll be , Chris!" . . . and I wheeled to look into the face of Joe Templeton, who, in his joy upon seeing me, had tried to knock a hole through my feeble carcass . . . "Stand still, you infernal simpleton, or I'll cut your hand off," said an attending medical officer as Joe danced about trying to give vent to his joy at my unexpected appearance.

Chapter 10

Trekking over a Battlefield

Colonel Cavanaugh, erect, quiet, immaculate, serious, appeared before the dressing station. He looked here and there at the many wounded on the ground, at the squatting stretcher-bearers now snatching a few seconds of rest at the end of their long journey off the field. Then a high-pitched voice, familiar to me, rang out: "With the colonel's permission, Sir, Sergeant Meeks will organize the ammunition detail." I looked around, and there he was. Hat askew on the side of his face . . . standing erect . . . a perfect military form . . . face immobile . . . eyes squarely meeting those of the colonel. "Very well, Sergeant . . . and perhaps that man will help . . . and this." (Pointing to Joe and me.) . . . "You're mighty damn tootin' they will, Sir" . . . "Where in the did you come from, corporal?"

After instructing Meeks to commandeer any and all soldiers needed by him to organize an ammunition-loading detail, to ration his men wherever he could find food, and to remain on duty until relieved, we left the colonel and went off down the hill just as the sun, wishing to have one look that day upon that blood-sodden scene, threw long black shadows into the valley before us.

Joe explained he had been cut on the hand by a wire covered with mustard gas and had come to the station to have his hand dressed. Then he laughed when he said he had been trying to 'dodge the detail', but now that Meeks and I were to be with him, "it ought to be a good safe war".

Meeks had spent the day at headquarters and had seen the fight as it went over the hill. He had learned of the necessity for the ammunition detail to load out trucks with extra ammunition for the men up front, and knowing the colonel was going to the dressing-station, had hastened there to seek the assignment.

The three of us went slowly to the bottom of the hill, hugging closely to the bank for protection against the 'big whiners' which still broke here and there throughout the area. Dusk was rapidly approaching. Men sat against the bank in groups. There was no excitement. A day's work had been done. Buddies met each other and chatted. No one seemed to be in a hurry. We came to a small pool of water. I wanted to drink . . . "You jibbering idiot, don't you know gas-shells fell all over this area today!" . . . I went on. Horses stood quietly tied to trees. As each convulsive blast from another shell ripped the air the horses, now accustomed to the sound, raised an ear in that direction and relaxed again. We headed into the woods, and it was soon black again with night's return.

Now thundering shells rolled across the skies and stabbed a crimson splotch against the inky night. Reacting to the eeriness of the shimmering shadows formed from the lurid flashes, the horses in the timber snorted and tramped their feet while placating soldiers called to their charges in soothing tones.

We stopped before a dug-out. A huge pile of ammunition lay nearby. I went inside and found several soldiers sprawled upon rudely constructed bunks. A pale yellow flame flickered from a candle stuck to a helmet. Outside again I looked about. A fire flashed in the timber some distance to the northwest . . . A kitchen . . . and I was hungry. Joe went with me. A new battery had moved in. They had hot coffee . . . delectable food. I sat against a tree and ate while we listened to the stories of the day. We would have liked to stay there always, to have drunk hot coffee, and eaten, and to have sprawled on the ground and slept, but we remembered our mission and retraced our steps to the ammunition dump. Joe's hand was hurting him, but we had duties to perform. The boys up front would be needing ammunition. This war must go on regardless of fatigue, hunger, and misery.

Ammunition carts, caissons,—horse-drawn, mule-drawn,—and motors arrived in a constant stream, and there were only eight tired men to do the loading. Box after box was lifted and pitched into place.

And the conveyance moved away into the night to be instantly replaced by another. We struggled with the weight of the boxes until far into the night, and then, to relieve ourselves at our task, we paired off two to the box. An officer, probably fresh from a snooze in a sound-proof dug-out, rode up on a horse and asked sharply: "What the hell's the matter here, sergeant? Can't you hasten this loading! One man to the box and make it snappy!" Sergeant Meeks presented himself before the interfering officer: "Sir, Sergeant Meeks speaking. I am in charge of this detail under specific instructions from Colonel Cavanaugh. These men, Sir, have been fighting all day . . . without the rest probably had by others . . . and I consider they are doing damn well!"

When the officer had spurred his horse away, a philosophical lad ventured the assertion that "there was about as much consistency in his request as there is between Jesus Christ and machine-guns."

The ammunition was piled into the last cart. I walked away to the dug-out where I flopped onto a vacant bunk . . . and when I awoke the sun was breaking through a cloud-laden sky. It was nine a. m.

Shaking myself back to a realization of the immediate past, but unable to understand what had happened, I was depressed by the bloody tragedy of yesterday, recollections of which smote my mind. I thought of myself in the language of Thomas Moore:

"Again therefore, like a sentinel of the dead, did I pace up and down amongst those tombs, contrasting mournfully the burning fever within my own veins with the cold quiet of those who were slumbering around."

With an effort I went outside. I could see a small clearing, a space of three or four acres, in the woods. There was a giant elephant-ear observation-balloon blobbing up and down on its picket-line. Around it for several hundred yards the earth was pock-marked every few yards with holes, many being as deep as six or eight feet. I was looking upon the result of the bombardment of the previous night, the holes made by shells which had fallen within thirty yards of me . . . and I had not heard a sound. Such had been the depth of my slumber!

I stumbled about through the edge of the timber looking over the night's destruction. I came upon the location of the kitchen where I had eaten the evening before. The kitchen had been moved. Retracing my steps in the direction of our quarters, I passed out into the clearing to stop before the huge balloons. I lingered long enough to see the men loose one of them. It rose several hundred feet into the air while the men in the underhanging basket made observations with glasses. The dim distant thunder of the guns continued to float back to me. The battle,—far to the north, to the east, and to the west,—was still going on, but with less intensity.

When I arrived at the ammunition dump a truck-driver was inquiring the way to the newly established dump at the front. I had not been there, of course, but having heard the drivers during the night receive their instructions as to their destination and having fought far in that direction the previous day, I felt I knew the way, and volunteered to go as a guide. Joe and Meeks heard me and both decided to come along. Jumping on top of the truck, sitting on thousands of charges of ammunition, and smoking cigarettes, we pulled out to the main road and swung around to the west toward Fey-en-Haye. Imagine our surprise! There, now, was no Fey-en-Haye. The only standing thing was one unmolested wall of stone. The other houses of the small village were crumbled heaps of gun-slashed fragments. Thousands of soldiers lined the road. The cry, "Give 'way to the right. Watch that truck", floated up to us as we jolted along the remnant of the road. Hole after hole in the road, many of which were several feet deep, was negotiated with the greatest difficulty, but we lumbered fitfully forward. Being held up for a short time before •entering this one-time village of France, we had an opportunity to view the surrounding country from this point of vantage, and to be able to understand something of the destruction which had befallen the area.

Our general direction was northward. The detour we had made to gain the road had placed us slightly southward and to the west of our location in the trenches before the battle. The timber which had once been green,—a deep, dark, approaching-winter-green,—now gnarled by the slash of explosives,—was losing its sheen. Huge trunks lay prone on the ground, their leaves fading from a rich green to a tinge of white. Limbs throughout the forest had been torn away, and heretofore graceful, spreading trees were transformed into stark, towering, branchless poles, through which our eyes now could penetrate to uncovered earth. Through these great holes in the forest the brown earth showed itself to us. Slightly to the rear of our position before the ruined village a road wound itself westward. Immediately in front of the road was a deserted trench, an endless brown jagged line across a descending plain which grew dimmer and dimmer as it traversed the valley to appear as a thin brownish stain on the low hill in the far distance. This trench had been the 'jumping off place' of the 360th infantry. The northward road onto which we turned gave evidence of having been at one time a substantial one. Now it was blasted, ragged, and pock-marked from the falling of thousands of high-explosives.

Soldiers,—members of the engineer corps,—by the hundreds lined the roadway toward the north. A stoical industry marked their movements as they loaded and unloaded wagons with stone from the crumbled town, and filled holes to make the surface of the road passable again. Germany, not even yet content with the carnage of the day before, kept her guns trained upon this artery which carried the life-blood to the boys up front, and shells popped phlegmatically all along the line of workmen. There was little contrast in the brown of the earth and the khaki-clad Americans, but spotted here and there could be seen the bright blue of the French uniform. They added touches of color and excitement to the scene.

The field to the right of the road, that area over which I had scrambled the day before, now presented a different appearance to me. For more than a mile to the north and east there was little evidence of motion. A chance soldier, usually walking with a cane toward the rear, could at times be seen, but the vista which presented itself was one of clay-brown destruction made ragged by the ever-present snarl of wire. Here and there an after-battle vandal picked over a corpse; some wore the olive-drab, others the blue. Far to the front a skirt of timber blotted out my view of details. This ragged forest of dark-green, with its familiar battle-riddled appearance, covered the area as far northward as I could see. Immediately in front of the timber-line the terrain sloped backwards toward me for about one-half mile, then it rose again to the very scene of the destroyed village. A small stream of water trickled through this valley, a stream only a few feet wide, a silver thread stretching westward glistening in the dull light of a gray day. I traced it with my eye far to the left where it wound northward and disappeared behind a jut of land, a hill covered with shell-torn trees.

In the valley to the fore, Frenchmen with blue coats, intensified with touches of red, busied themselves with their teams. They were dragging huge cannon, and their teams lurched and charged with the report of each shell yet falling with methodic rhythm. Passage across this undulating plain was impossible for teams, except in the roadway. And men walked there with difficulty. Trenches,—long winding lines of trenches,—gashed the whole face of the earth. Barbed wire,—snarled, stretched tight, tangled, browned with the decadence of the elements,—warned you against passage, foreboding evil consequences to him who tried.

Our truck moved forward almost to a cross-street near the crumbled piles of stone. Here again we encountered the traffic jam. A truck had lurched cross-wise in the road, and dozens of waiting trucks had augmented the confusion. Legthargic soldiers, some grinning, some solemn, gingerly lifted huge cannon shells from waiting trucks and piled them row on top of row by the side of the road. The bright yellowish brass-jackets gleamed a streak of light marking, as it were, a passageway through the scene of devastation.

Again we moved only to stop before the only standing wall of the ghost-village. An American officer, vituperative and loud, approached a crowd of dough-boys lying against the tottering wall. His repeated orders only caused them to move and then with a reluctant tread as if to say: "What's the difference how you get killed,—just so you do?"

Clearing our truck from the concentrating traffic, our driver slipped forward and pushed past slower moving vehicles and crowded walking soldiers into the water-soaked ditches, and when we had gone beyond the limits of the village we bumped along the road which had been newly covered with the broken stones. A cordon of engineers stationed at short intervals continued to unload stone into the roadway while others sat, and with great hammers rapped the stones to smoothness.

Amid the rap . . . rap . . . rap of the stone hammers, a voice came up to me: "Hell-fire and damnation, Emmett". I looked down to the right and there was 'Doby' Lloyd. I called to the driver to stop, jumping simultaneously with the call. An on-coming truck forced our driver to the right and he skidded into the ditch from which he was unable to extricate the truck without assistance . . . But I put my arms around old 'Doby', and how glad I was to see him! In one unremitting breath he would have told me the full story of the day before and his part in the Big Fight. I broke into his story to inquire about Goodson, Singleton, and Rogers Tipton. When I spoke of Tip, Doby turned toward the east and pointed: "Just on the top of the hill. I've seen him", and when he looked at me again tears were standing in his eyes.

Another truck hitched onto ours and we were yanked back into the roadway. The rumble of our truck floated back to Doby as he stood by the roadside waving his steel-'lid' after us.

The many passing vehicles had converted the trickle of water across the roadway into a veritable quagmire. With a rush our truck slithered through it, swaying and bumping while we sat on top of the load laughing at the fretting 'Frogs' who whipped their teams in a frantic effort to haul their cannon farther down the valley.

I looked to our right at a cross-road. Sixteen horses lay dead. A decimated cannon, broken wheels, distorted gear, fragments of men with blue-uniforms, showed that a direct hit by a G. I. can had occurred. Up the hill we went and the shells were again breaking all around us.

Then I saw something,—a pointed steel-helmet! I cried out to the driver, motioning in that direction while he brought the truck to a stop. I slid to the ground with my rifle trained on the helmet-point which was bobbing up, then disappearing, in a trench some two hundred yards away. "Don't shoot", shouted Meeks . . . "It's just a lousy souvenir-gathering-Frog", which it proved to be when two Frenchmen wearing salvaged German helmets crawled into view. "No wonder the Frogs can't win a war!"

Once in the woods our progress was rapid, the road there not having suffered the effects of the bombardment. When we made our exit from the timber we were on the edge of a very steep hill. American soldiers lined the rim. The valley below was filled with a greyish smoke. Shells were falling at every road-crossing. Across a comparatively narrow valley was a hill some three hundred feet to its crest. It was covered with a shamble of timber about halfway down from the top. Between us and the hill and beyond a shimmering stream which bore off down the valley to our right lay the little French village, Vilcey. It seemed to press against the hillside for protection while its life- blood, now running white, slowly ebbed away. Trudging in broken lines American soldiers were going forward. A renewed bombardment into the valley was Germany's warning that death lurked there. To the left of the town, cross upon cross, row upon row, white-markers pointed out the last resting place of many of Germany's dead, the result of a carnage which had gone before us. A whimpering, foot-sore, weary soldier, burdened with an over-coat and bearing a rifle squeezed up under his arm, fell out from the on-flowing ranks to lean against our truck now momentarily stopped on the brink of the hill. A big 'Jerry' whined out of the skies and crashed into the cemetery. A coffin ruthlessly routed from its resting place flew high into the air, and my new found friend whimpered:" ---it! They won't even let their own dead rest."

Once the road was clear below us we dashed down the slope, rounded a curve on two wheels, slushed through the fast moving stream, and entered the village. Here was pandemonium. Shells were crashing all over the place. House-tops toppled in, and soldiers darted across the streets only to escape the crunching wheels of speeding trucks as if by magic. The flow of vehicles and men seemed to be up the main street of the village at the head of which traffic was congesting before a building. I was curious and slid off the truck, entering the foremost stone building in and out of which poured a constant stream of khaki clad boys. Behold! What a sight! An American Armistice had been declared! The fleeing Germans had left behind a big boiler of steaming stew. A German bar, well stocked with beer, had been too bulky for evacuation . . . and Young America was in charge! Dripping ladels of German-stew dumped into extended mess-kits, and bottles popped to the tune of:

"'Twas a long, long trail a winding,
Across all France up to here:
But it's lucky we're afindin',
This fine old lager beer."

"Whew . . . ew . . . ew . . . Wham . . . Bang" . . . and a shell broke just to the rear and over the 'oasis'. Soldiers scattered in all directions, some running with foaming bottles, others with dripping mess-kits. I went through the kitchen and out the front. Two more shells landed short of the town, spraying shrapnel against the house under the eaves of which I stood . . . and I knew this was not the most comfortable place in which to linger.

Returning by a circuitous route around the house, I encountered an officer who was literally frothing at the mouth and demanding that "Whoever in the hell is the guide for this truck, get it out of here." Being convinced that he probably had me in mind, I mounted it again and we rumbled out into the open before the town. We circled the base of the hill which we had seen from the distance and crossed a road which led eastward into the valley. I spied a camouflage road-covering ahead and pointed it out to the driver. We stopped behind it, and much to my surprise there was the ammunition dump for which we sought. A hasty unloading put us back again on the roadway and we clattered back again at a terrific pace.

Negotiating the road-intersection where the shells continued to fall, and escaping a burst only because it fell a little short, our highly energetic driver swerved around the corner of the 'kitchen' with shrieking brakes. He too had hopes there might be 'convivial remains'! But to our sorrow the boys had licked the platter!

A "Medico's" car spurted into town and jammed into the back-end of our truck. Two lieutenants alighted, one bleeding from a fresh wound in the arm. The other's face had the pallor of death. They told us their story . . . Up the hill to the north of where we had located the ammunition dump they had run afoul of a machine-gun nest. The 'wasps' had stung them all over. The car, along its full side, had a line of bullet-punctures,—a little too low for fatal results!

"Bam!" . . . One, . . . two . . . three . . . four . . . they came dropping into the village again. The lieutenants ran into the stone-building, but my driver was grinding his gear. Down the street we went . . . with klaxon rasping out its raucous call of warning . . . through the slush of the stream. And I turned squeamishly to look back upon the valley, for I now realized we must climb the hill. With the truck's speed at a maximum . . . climbing slowly . . . always slower . . . now in second-gear . . . now in first . . . we were a perfect target plastered against the side of the mountain! Shells whined and broke. The driver swore and pressed harder on the gas, but the truck only whined . . . and climbed . . . until it gained a curve in the road. Here it picked up speed and mounted to the top. Darting again into the timber, I felt a sense of relief, and banteringly said to Meeks: "Think I'll take this job permanently, Meeks!" . . . "Well, you blithering idiot, you can have it . . . If I'm going to fight I want to see something to fight . . . G. I. cans . . . whizz-bangs . . . and what-the-hell-have-you don't suit me."

We came to the timber's edge again and looked upon the scene of yesterday, splotched with innumerable, brown shell-holes, a pitted earth suffering from man's greatest crime,—War! As I looked upon this scene of desolation,—yesterday,—a night-mare of exertion, mental distress, incomprehension,—floated through my mind, again, like an ominous cloud. Before me were the results of yesterday. And as the shells continued to rumble out their devastation, I thought of how Shelley must have felt when he wrote: "No change, no pause, no hope, yet I endure."

I left the truck. Meeks called after me: "Where in the name of all that's holy do you think you're going, now?" . . . "Come on, Joe . . . Shake a leg, Meeks . . . Let's take the shortcut back . . . Let's see!"

They came, and across the blood-sodden field we walked. Kaleidoscopically the scenes of yesterday crowded back through my mind:

"a big mortar-shell wobbling with its energy almost spent, clearing our heads only by feet. Instinctively I turned to watch its progress. It thudded into the soft earth a few yards to the rear. From it oozed a thin, green smoke. Just beyond the fallen shell, and coming toward the salubrious gas, was Clarence. He was running toward me unmasked. The characteristic grin was on his face. Suddenly he stopped. . . . Wavering a moment . . . weaving . . . he went to his knees . . . over on his face."

Was this just a dream of yesterday? I would see! The road . . . Fey-en-Haye, that ragged splotch against a grayish sky . . . served to point out our direction, and we turned from the route of travel . . . to the left . . . over wires . . . into deep spike-infested trenches . . . to the hill to which 'Doby' had pointed: "Just on the top of the hill I" . . . We came upon familiar ground . . . "It must have been just beyond, Meeks." . . . Two men,—khaki clad, with iron-hats, tattered cream-yellow raincoats slit into tatters by clinging wires,—now slogged slowly toward us. They carried a stretcher. One called out: "See any 'stiffs' over here, fellows?" . . . "We're looking for Clarence Oliver, now . . . Ought to be about here . . . Yes, there's some lucky devil over there. See?" . . . A clean expanse of reddish mud . . . a dark brown form . . . and Meeks was standing erect with steel-helmet in hand. . . . He was looking straight ahead . . . He seemed to be peering into the future . . . The muscles of his face quivered . . . Clear blue eyes were moist.

I laid my hat upon the ground. The stretcher-bearers approached. "Leave that boy alone, fellows, just a moment, please." A contracted leg prevented me from turning him upon his back. Reverently, Meeks, then Joe, caught the stiffened form, and we turned him from his face. It was Clarence with the same cheerful smile. He had carried it into Eternity! "He was not hit, boys . . . I saw it happen . . . It was gas." . . . Tenderly, carefully, we looked . . . There was not a mark on him . . . And they bore him off toward the west!

"Tip should be over there, fellows", and heavy hearted we turned to the right and up the hill. A long gangling form emerged from a brambled trench,—Old Bill Goodson. . . . "Bill, we're hunting for Tip . . . Doby said he was over here." . . . "He was, but they took him away . . . Shot through the heart."

Bill extended me the broken stock of a rifle. "What you lugging that thing for?" . . . "A little souvenir . . . Last night . . . all night . . . I stood in the mouth of a dug-out waiting for the counter-attack . . . A shell-fragment struck my rifle . . . This's what's left . . . Will make a good souvenir if I can get it home" . . . "Home? . . . Think you're going home? . . . Ha! Ha!"

"But where's the company?"

"About four miles up across the second hill".

"How did the boys fare?"

"Not so good. Thirty-one failed to report this morning. You are one of them. Captain is reporting you lost in action."

"Correct the impression,— if he's really personally concerned about me."

"Old Molly's gone . . . Shot through the head . . . I was with him when he was hit . . . Machine gun took away his whole temple . . . God! I'm glad you didn't see how he suffered . . . Do you remember Molly and his father at Texarkana?"

The drab green of the timber behind invited us under cover from the shells whining over the field in search of someone to maim. We sought out the medical field-station, for Joe's hand cut by the merciless rusty wire had developed stabbing pains. We walked down the hill where the evening before we had met each other. Now the relief station was abandoned. Blood-soaked gauze was scattered upon the ground. Our tramping feet scared up a maze of flies which buzzed about our faces before they settled back to feast upon some man's blood.

"Hi there soldier! Going to headquarters?" . . . A be-spattered lad with rifle in hand and bayonet fixed came punching a German soldier ahead of him from out the bushes . . . "Wish you'd take this 'Kraut' with you . . .

Found him sneaking back toward the Vaterland . . . Thought I'd pot him but he kameraded, then decided to bring him in . . . Maybe he can talk."

We relieved the soldier of his charge, who seemed to be anxious to wander back toward his company at the front, and we marched the Heine along ahead of us toward the brigade headquarters. I tried to talk with him in English, then switched to German,— the little I knew,— but he was sullen and uttered only gutturals which meant little to me.

An officer met us at the old ammunition dump. He demanded we deliver our charge to him then and there. The glint in his eye conveyed to us his hopes that he would be accredited with having captured a prisoner. Before leaving us, however, he requested I make another effort to talk with young Heine. I elicited the information that he was a Prussian; that we had fought against some of the Prussian Guards the day before. He also said that the ceaseless activities of the days before in our front had apprised the German command of the impending battle. Preparations, therefore, had been made to let us spend our force against an evacuating field. The German army had attempted to retreat to the Moselle River and dig-in. The plan to leave the field occupied with machine-gun emplacements had been carried out, and the Germans left there with these guns had been designedly sacrificed for the safety of the greater part of the army. . . . And with this information the gallant American officer marched off with 'his' captive!

The trucks were waiting for us at the dump when we arrived. Again we climbed aboard and struck for the front with adventure little different from the trip just completed. On our return trip, however, we saw the roads were being severely shelled with a determined fierceness from the German six inch guns, so the three of us left the truck again with the intention of making our way across country, leaving the drivers to care for themselves the best they could. Why should we add consolation with our presence? Hadn't these men gone into the truck service because it was a 'snap'?

My legging came down as we trudged along and I stopped to wrap it. The others went ahead. A car drove up by my side and stopped. It was occupied by three officers and a driver . . . "Is the road clear through the woods?" . . . "The road's clear all right, but nobody but a cock-eyed fool would try it through that shell-fire unless he had to go." Two officers sat up straight, then . . . General Pershing leaned forward and smiled: "Thank you corporal for the information . . . and the compliment!" . . . I stood in the road and looked after the disappearing brown car . . . "Game old rooster anyway!"[1]

I now could see men working ahead of me and went to them. At first I could not make it out,—then it was plain,—a rick of dead American soldiers piled like cordwood five or six feet high and thirty feet long . . . Knowing that Tip and Clarence and Molly were there, I shambled off toward the distant timber again which was now fast losing its distinctiveness with the graying of the day. Then I decided to ride, and swinging a truck clattering along the road, it took me back to Fey-en-Haye and toward the original 'jumping off place'.

At the cross-road an ingenious Frog . . . (or perhaps it was a venturesome Y. M. C. A. man . . . either one would take your money,) was displaying *New York Times* (printed in Paris) for sale. Instinctively, following the habit of a life-time, I jumped from the vehicle, and, snatching a paper, sat at the corner of the delapidated dressing station to read the news. A gangling American machine-gunner approached and stood over me. "All right, Big Boy," I said, "you can have it next, but don't crowd me. I want to read it." . . . But when he laughed I looked at him. He was my old friend, Miller, a friend of boy-hood days, whom I had probably not seen in fifteen years. We began to speculate on the smallness of the world, and the coincident that we should meet here after so many years of separation, but Heine would not have it so. He drove me into the woods for protection, whilst Miller went his way.[2]

---

1 (Ed.) Combat troops on both side wore wool bands wrapped tightly in a spiral from their ankles up to their knees to keep the trousers tight around their shins and calves. The wool kept them warm from body heat even when wet, the wraps made it more difficult to catch the cuffs on barbed wire, and the tight seal prevented mustard gas and other blister agents from creeping up their legs.

2 (Ed.) Many of the Doughboys developed a hatred for the YMCA in France because the men who worked for the organization were exempted from service, wore a uniform that resembled that of an officer, and they charged full price for all the comfort goods that were sent overseas from cash or goods donations from the folks back home. They were compared to the Red Cross and other voluntary organizations that provided donuts, cigarettes, sandwiches, coffee, stationary, etc., free of charge to those in the military in , hospitals, stateside training camps or from canteens located just a few miles behind the front lines.

Night's arrival brought many more trucks to be loaded with ammunition. We worked steadily, wearily along. A messenger came, advising that the 357th Infantry was to attack at daylight and saying we had, by mistake, loaded out blank hand-grenades for this company's use. What a pickle? Meeks fretted around trying to commandeer a truck, a horse-drawn wagon, a mule, anything, but all were working under orders. "We must get loaded grenades to those boys before they attack." . . . "Sergeant, if it's all the same to you, I'll take three boxes of grenades to those boys" . . . "The hell you will! You spider-legged blob of Irish enthusiasm, you can't walk that far to say nothing about carrying three hundred pounds of grenades." . . "How I'll do it, Old Top, will be my problem. Give me three men . . . There are now eight of us." . . . "O.K. and Good Luck . . . You know the way . . . Down the road to the east through Ponta Mousson . . over the bridge to the left . . . up the mountain to the top . . . and report to Captain ---."

Out of the woods through the impenetrable darkness the four of us made our way chiefly by feeling with our feet. The old plank corrugated-walkway served as a guide. This trail, unused by me since the night we entered the area, seemed like familiar ground to my feet. When we came to the road our course was downhill, but the steady weight of the heavy boxes slowed down our pace. Our weary bodies required many intervals of rest. Trucks ground past us, speeding along without light. Every effort to halt one to elicit a 'lift' met with equal results,—nil! Finally a sound wafted to us . . . "knock . . . knock" . . . the dull flump of slowly revolving wagon wheels. A two wheel bread-cart being drawn by a diminutive jack-ass came abreast of us. I grabbed his bit and pulled him to the side of the road while the driver cursed through my explanations.

The driver and his Bucephalus were going to Ponta Mousson with a load of bread, and if we could stack on the grenade-boxes the jack-ass could pull the load down-hill . . . And away we went, shoving the commandeered cart when a rise in the road was encountered and resting after pushing when the cart rolled free.

We saw a light flicker under a thinly veiled tree-line some twenty-five yards to our right. Then men began to run . . . Frenchmen called out into the darkness: "Voila! . . . Voila!" . . . A crimson tongue of fire reached out toward us, just over our heads and . . . "Blam!" We staggered from the concussion, the full force of which had been shot straight at us . . . "Oeil de Boeuf!" (A bull's eye) We had been in the direct line of fire of a twenty-one inch naval gun.

At the bridge a sentry stopped us. We made ourselves known and passed on beyond head-quarters where dim lights showed through curtained windows. Men were coming and going. Officers with bespattered uniforms left hurriedly into the darkness. There was a grimness about the manner of these men. Other officers discernible even in the meagre light for their sprightliness entered the quarters not to come out again,—bomb-proof brass hats!

When we came to our turn to the left, Medusa's colt and his driver left us. The bread-cart flumped down the cobbled street. As we again shouldered our burdens the lines of Shakespeare came back to me: "I was not made a horse; and yet I bear a burthen like an ass."

Up . . . up . . . up the long, steep hill, step by step, faint and weary, we made our way always watching for signs of 'dug-in' soldiers.

"Captain ---, 357th Infantry, Sir?" . . . A detail with grenades."

"Sergeant take these grenades and have these same men take the blank ones back to their dump!"

Upon hearing this, my blood boiled. We lay on the ground for a brief rest while the sergeant returned with the boxes of blanks. We picked them up and strode ten yards into the darkness. I kicked Joe, and we placed the boxes against a convenient tree. There we left them. They may be there yet so far as that detail is concerned . . . And at two o'clock in the morning we reported again to Sergeant Meeks!

Eight big trucks were at the ammunition dump when we arrived. Meeks and his four men were struggling with the work. Weary as we were our aid was limited. Joe crawled into a truck to help the driver place the boxes. Then he stepped back onto the ground and whispered: "Who is that driver? Voice sounds familiar" . . . "Big Boy, what's your name?" . . . "Jim Alexander." . . . "Old Jim . . . Old Jim", we shouted in our joy, and Meeks swore he would shoot the whole foolish gang before daylight if we didn't quit making so much noise. Even then the Germans had their aviators purring around in the sky above us trying to locate our position, now mercifully obscured by the night and a gathering fog.

Vilcey, France, in "Death's Valley" (note shell-hole in church-spire)

Eight trucks loaded with high explosives, thirty-two boxes weighing one hundred pounds each, running bumper to tail, ground out onto the road toward the front. We had found Jim, so Joe and I, despite the danger to be encountered in making the trip, mounted the canvas-covering just to the back of the driver, and went with him. He drove the third truck from the front. Through Fey-en-Haye, escaping the snare of the valley-quagmire, into the blackness of the woods, and tipping over the hill for the descent into the valley, we rode talking to 'Old' Jim. Then he bumped his truck into the one ahead. It tilted treacherously, then slumped into the ditch where it stuck fast in the mud, effectively blocking the road behind it. A big shell cracked over our heads. American engineers alongside the roadway, stringing wire-fence . . . ever and ever . . . more and more . . .

wire-fence . . . cried out: "Run! Your truck's on fire!" Joe jumped to the ground. His foot caught in the canvas which precipitated him head-first. There was a dull thud and a groan. Following quickly I fell on my back directly under the wheels of the truck. And there above me the underside of the bed of the truck was ablaze. Instinctively I began digging into the rough gravel-road. The flinty stones cut my finger. I collected handful after handful of moist dirt which I threw and rubbed on the flames . . . And the fire went out. Crawling from under the truck, dimly I saw Joe raising himself to his knees, groaning. I grabbed him. Another shell rent the air. Shrapnel tore into the dirt around us. With an arm around Joe, half running, half dragging him, we sought the shelter of a nearby bank . . . He had landed on his face and had fallen against his steel helmet. Blood was running from a mashed lip, but he spat and laughed: "Paint my with iodine and mark me 'Duty.'"

Reassembling, we unloaded the truck, cleared the roadway, and eased down into the valley to Villers-sur-Preny. Here the seventh truck broke a wheel while crossing the stream and turned partly over. Wading in the cold water, we transferred the ammunition hastily to another conveyance, threw the truck on its side in the clear of the road, and dashed ahead through the darkness to the ammunition dump. Here we were greeted with a salvo of shrapnel-throwing shells some of which struck boxes on the dump. Finishing our work by herculean efforts we beat an unhesitating retreat toward the rear. A milky mist had now enveloped the earth. The grind of many gears and the whir of wheels had attracted daring aviators. The purr of their motors followed above us. We chased along hoping to escape them. Nearer and nearer they came. The wobbly purr above told us they tried to locate our position. A flare was dropped . . . Down . . . down . . . it came, lighting us up from our milky obscurity . . . "Crash! . . . crash"! They dropped bombs after us. They missed by fifteen yards . . . and we gained the timber . . . and safety!

With daylight orders came to move ahead a mile to another ammunition-deposit. We packed up our scant belongings, and sensing the security of the timber one mile ahead, left on foot. Under over-spreading trees with boughs piled high on the ground we found our destination. We looked about for a 'rat-hole' in which to live, some place for seclusion. Meeks pulled back a pile of yellowing brush and a man's head popped out of the concealed hole in the ground . . . "Hi! Big Boy, pretty soft, eh?" . . . "A regular cave, fellows . . . deep and long . . . sound proof . . . bomb proof . . . plenty of room . . . my home until the "end of the emergency" . . . Want to share it with me?"

Entering the mouth of the cave, I counted . . . down . . . down . . . Thirty-one steps had been cut into the soft stone. At the bottom we were in the bowels of the earth. Dank darkness . . . rancid odors permeated the foul air. Another fellow came along the narrow alley-way. "I'm leaving boys . . . M. P's. ordered me 'out' and 'up' . . . Candles in a room to the left and straight ahead a hundred feet. Damn lousy place if you want my opinion about it."

Feeling along the sides and bearing to the left as directed, the mucous scum splattered from the walls sliming our faces and hands. There was a sharp turn to the left. There was a table. Someone produced a match, and the candles were there!

Our light now showed a room cut in the soft rock. It was some fifteen feet square. Here was a veritable work-shop of iniquity. Shell-box after shell-box lined the wall, each shell having a cankerous tip, the result of having been dipped in a green powdery-poison. German's hell factory, indeed!

A sound, . . . low, mournful, . . . a wheezy whine, . . . reverberated through the hallway. I took a candle and sought out the source. I passed through cavern after cavern. Then the sound was nearer. I found a boy lying on a rudely constructed bunk. He was crying . . . low . . . wheezily. I held the candle over him. Watery, scared eyes looked up pleadingly to me. "It's all right, old fellow. What seems to be eating on you?" . . . "Where am I . . . The gas . . . O, my chest." . . . And as I listened to his labored breathing I knew he had pneumonia probably brought on by gas inhalations.

An M. P. came and we made him as comfortable as possible, giving him water and a more comfortable bed . . . but when the day had waned, he had gone!

Joe's hand had become fiery with infection. He must have a doctor. A truck was going back to a repair base some ten miles. Emptying my pocket of the francs I had, and with . . ." Watch it . . . Don't get into the hospital", I saw the retreating truck bear him toward the rear.

A wheel-kitchen moved up close to our new dump. Meeks protested violently: "Your fire will attract the Big Ones," but it was of no avail. They too had orders to stop here. So we compromised by having the cooks prepare us a supper and we ate with them.

I leaned against a tree and looked eastward. The towering top of the mountain above Ponta Mousson . . . that hill of a slothful jack-ass and a wretched night for Joe and me . . . turned with the last receding rays of the sun from a shimmering pink to a huge, greenish-black cone, pointing silently heavenward. I was lonesome. Joe was gone. I knew not where. Meeks was away seeking a less slimy, lousy bed . . . and I was alone with my thoughts . . . deep in the forest, its fast enmeshing blackness stifling me . . . "Whew . . . ew . . . ew" came the drone of a 'high explosive'. Electrified by its piercing fierceness, and realizing it was headed straight for us, I grabbed a tree-trunk. Then I ran behind the wheels of the kitchen and squatted as if for protection upon the ground. "Wham" . . . it broke over the dump . . . and a sickening whine struck a trajectory across the sky above. Slashing off a limb from the tree from which I had just moved, a slug of steel buried itself in the ground in the footprints I had just made! With a convenient hoe we dug the sizzling-hot shrapnel from the ground. It had sunk two feet deep and weighed about four pounds.

I walked away into the accumulating darkness only to find that Meeks had located a corrugated-metal-covered dug-out. In it we lay down with many soldiers. The shelling continued, each one falling closer and closer to our quarters. I aroused Meeks and we walked back to the cavern and slept on the table in the 'poison room'.

No night was spent without the protection of a gas-guard. Our guard called me in the early morning for my 'stand-to'. Whining shells even then were playing over the area. I heard shrieking human voices raised in the direction of the corrugated dug-out which we had left. When morning came we were lured there by the cries for help. What a mess! A half hundred men were lying on the ground, some in the dug-out, some on the adjacent ground. Some were dead. Others were writhing in the agonies of mustard-gas burns. A tall, slim American officer was directing the work of rescue and palliation. His was a familiar face, but there was no time for acquaintances here. But when he departed,—too far to recall,—I remembered. He was Bland Proctor.[3]

Our now more advanced position made it easier for us to let the truck back-track along the road through Fey-en-Haye whilst we cut straight across under cover of the timber. By doing this we could catch up with the truck with only a short walk, and the walk permitted us to see much of the area over which the armies had fought. While on one of these walks on the south hill before Villers-sur-Preny we found a maze of caves, impregnable from attack from the south. These holes were capable of accommodating thousands of men. Quarters were neatly equipped with tables, chairs, electric lights, and an occasional piano. Evidently the Germans had intended to stay here, but Burns knew the ways of the world when he said:—

"In proving foresight may be vain:
The best laid schemes o' mice an' men
Gang aft agley,
An' lea'e us nought but grief an' pain,
For promis'd joy."

Meeks and I left our explorations into the hill-side one day to venture a direct course across the valley. We forded the stream of clear, running water, passed through the velvety-green of winter-grass growing in the watered low-land, and ascended to the barren brown-clay slopes beyond. It was our intention to strike into the timber and skirt it to the unloading station. Half way up the brown expanse we espied an airplane. We, too, had been seen, and he spiraled out of the sky to come down upon us. We ran,

3 Of Victoria, Texas.

gaining a shell-hole. Crawling from one hole to another we tried to conceal our position, whilst the aviator whipped over trying to locate us. We lay upon our backs, plastered against a soil the color of our uniforms, and running the brown mud over our faces for disguise (cutting off the shine of the skin), we watched him. Here and there. To and fro, he played, now low, then higher, seeking, searching. Spiraling again to gain altitude, he paused high above the timber-line to the south across the valley. A signal,—puffs of smoke,—came from his plane, and from far off in the direction of Metz a whine came toward us. One . . . two . . . three . . . four . . . the big shells whistled across the cerulean sky . . . "Wham . . . wham . . . bam . . . bam", they exploded with terrific concussions . . . And Jerry had made a direct hit upon an American battery concealed across the valley from us.

The aviator, now, having done his bit in that locality, drifted westward. We followed in his wake. When we neared the ammunition dump, we caught up with and loaded onto one of our trucks. It was a small, chain-driven Ford of uncertain vintage and doubtful efficacy, which broke into a paroxysm of metallic chortling when the driver tried to force it along with its added burden to our nearby destination. The mellifluent 'Ford-song' floated out through the valley to attract the ears of a passing enemy-plane. A priest, methodically administering the last rites over a multitude of khaki-clothed forms, called out to us: "Why in the name of the most Holy Spirit can't you boys get away from here with that thing and let me alone". As I looked at him he went back to his work of entombment, at the same time, apparently, oblivious to the approach of the German aviator whose wobbly motor was now directly over and turning upon its nose to make a dive at us. A camouflage fence stood to the left. I scurried to it and crawled along on the ground as the bullets rattled down on the truck and swept the stakes above my head. What a narrow escape we had had!

He gained altitude for another try at us. I took advantage of his rise and ran to the right into the timber and fell behind a tree. Meeks peered from behind another: "You're the slowest sloth in all France." But Jerry had seen where we went. Again and again he whipped over the trees to throw burst after burst of machine-gun bullets into the area. We retaliated by emptying several clips from our rifles into the plane. Perhaps we had made it uncomfortable for him, for he rose and released four bombs. They splashed destruction around us, uprooting trees but failing to harm either of us. One bomb was a dud.[4]

When the aviator departed, our crippled, asthmatic Ford clanked chatteringly off down the valley. This time we were headed for the nearest approach to Pagny-on-the-Moselle. Half-way up the mountain-side the chain-drive, just previously struck by a bullet from the intrepid aviator's gun, rattled, lurched and jumped its gear. We were stalled and our tool-box contained a lone screwdriver,—nothing more!

The swearing, frothing, fuming driver would 'leave her here' and 'to hell with her and all her kind, including its peace-loving maker. But our louse-infested hole was more than four miles away; the day was drawing to a close and we were weary. I would not consent to pushing her into a convenient limbo but set about the task of making her run again. Laboriously loosening a bolt from its fastening in the wood-bed of the car, and straining with fingers and teeth, I succeeded in splicing the chain together. Then a big shell came whirring over-head and clipped the top of the tree beside us. Down came a shower of twigs and leaves, but we were too close to the mountain's breast to be touched.

We feared to place too great a strain on the point of my improvised repairs, so we walked the remaining distance to the top of the mountain and notified the officers of the location of the surplus ammunition. Retracing our steps we boarded the old Ford again and she rolled back into the valley through a gas-filled village which lay at the foot of the mountain. A newly found officer, seeking a respite from the front by 'hitch-hiking' with us, begged for more speed through the village or "we'll strangle to death". We rolled out into the unencumbered valley but paused long enough to roll the dead form of M Company's cook from the tread of our wheels. His field-kitchen, demolished in the ditch, his blood slowly congealing from gaping wounds told the story of the roadway of the half-hour before. There

4 (Ed.) The "clips" were tin brass expendable clamps that held five rifle cartridges each. These five rounds filled the magazines inside the rifles carried by the Soldiers. When all five were fired, he had expended a whole clip and it was time to reload with another.

he was with fixed, staring eyes, a bruised and bloody face partly concealed with boyish whiskers . . . I tried to reconstruct him into the genial chap of the days of Colmiers-La-Haut when I was hungry and he fed me!

"'T was not a life,
'T was a piece of childhood thrown away!"

C est le Guerre!

Day after day the rain trickled out of a lowering sky. The hanging branches of the trees splashed us. Waist-high bushes rubbed off their accumulated dampness . . . We were wet . . . We were cold . . . Night after night we lay in the infested cavern only to arise and seek out a flickering candle and run the seams of a louse-infested shirt over the pale-yellow blaze . . . The pungent, acrimonious stench of frying vermin bit our nostrils . . . And Old Meeks' high-pitched, encouraging, jovial tones would come across the table to us: "Well, what's the news from the front? . . . I see you're reading your shirt again."[5]

L. C. (Elsie) Barron, mess-sergeant, the keeper of the comfort of the bellies of L Company, extraordinarily haggard, worn and distraught, returning from a trip to the rear in search of supplies, shambled past our retreat just as the day was waning. What a joyous apparition! . . . He had fed us . . . Now we took him in!

He brought the news. 'Red' Crow was dead. Captain Whitaker was acting major . . . Lieutenant Nysewander was in charge of the company . . . Lieutenant James A. Baker had returned from the training area and was a welcome addition to the company . . . Lieutenant Montgomery Fly had fallen on the field, bullet pierced . . . He had refused evacuation as long as others needed attention . . . Hour after hour he had lain there . . . day . . . a night . . . a day . . . and they carried him away . . . And he died in the hospital . . . The boys had dug-in before the Moselle. . . . Sergeant Meeks was to report to the colonel the following day . . . But Elsie was so tired and

5 (Ed.) "Reading a shirt" was an ancient occupation for old Soldiers who removed their wool garments any time they had an opportunity to pick and destroy lice that would bite them mercilessly. By running the seams of their clothes over a lit candle, they could kill the lice and their eggs while taking care not to scorch the cloth.

we let him sleep . . . and at the first signs of returning light, I shook him back to realities . . . A sip of cold coffee from a clanking canteen . . . all we had . . . and a shambling form again passed northward into the bushes. He was our first contact with the company for many, long strenuous days.

A truck reported for ammunition, and Meeks decided to go with it to headquarters. I rode along as far as the advance-dump. Meeks was to meet me there later . . . A long, long, wait, and crunching feet came off the hill-side. Bleeding, limping, they came single file and slowly . . . Did I have water? They were hungry . . .

Against the mountain-side I had espied activities indicative of a concealed kitchen, and I led the way. Close up against the breast of the mountain and piled high with brush, a rolling kitchen gave out the succulent aroma of boiling food . . . And "You are welcome, boys, but keep under cover . . . Jerry has been reaching for this all day." . . . Mess-kits tinkled menacingly as the genial sergeant poured again and again. Food . . . hot . . . invigorating . . . made them cast aside precaution . . . Jovial banter floated up and up . . . And they told of the night before. . . . Of the gas-barrage . . . and how they were wounded. . . . And now they were ordered to get back as best they could . . . A lazy plane, with motor cut silent and black crosses showing, tipped over the hill . . . Then it fought valiantly for altitude . . . Puff . . . Puff . . . and the position-signal was released. A sharp whine almost immediately came down the hill from the northwest. It struck with a splash in the water at the base of the mountain. Jumping to our feet we dived headfirst into a corrugated-iron covered pit. Another shrill whine and the earth shook. The covering from over our heads rose into the air, and I was sitting looking up into a cloud-flecked sky. The shell had ripped away the roof from our heads. The kitchen was over-turned and the soup was spilled on the ground.

Meeks arrived and we wandered southward again. We skirted the timber-line to the fore of Villers-sur-Preny while the gas-shells thumped down with unfailing regularity at the cross-roads. A lone rider cantered up the valley in our direction. His sleek horse flashed silver glints with every movement. A horse and rider was rather unusual in this sector. And we knew he was "a bomb-proof officer" or a "stay-behind". We watched them as they came closer and closer. Meeks pointed out the horse's reddish-brown coat of hair as convincing proof of his being a "bomb-proof officer". Then he mumbled out his disgust: "That's him . . . The same parlor-soldier I had the round with the first night we loaded ammunition . . . And he takes charge tomorrow . . . Orders are to disband my detail tonight . . . Thank God, you can go to your company . . . You don't have to stay with that officer." . . . Then dear old Meeks looked at me with a cynical smile and added: "He'll probably get a croix de guerre . . . 'Valiant service rendered transferring ammunition under shell-fire' . . . But what will you and I get,—if we live through? . . C est la Guerre, eh?"

The horseman wearing a regulation khaki-hat, with helmet strapped to his saddle, pitched forward onto the pommel of his saddle when his horse scented drifting gas and shifted his direction toward higher land . . . "I see now. He's the sissy from Belton. God pity the boys, if they need more 'fire-works.'"

Preferring not to meet him, we scrambled higher up and walked along the ledges and passed above the road. An early winter-wind pushed steadily off the mountain-tip before Villers-sur-Preny driving the pale-yellow gas-laden air of the valley away from our course . . . "Looks bad down there this evening . . . Let's climb around this mountain point and come out to the right of Villers near the Boche cemetery" . . . And we scrambled along . . .

A gurgling sound came from above me . . . Some one called: "Gas." I sniffed the air and failing to smell anything, continued to climb . . . Then my nostrils tingled. A stinging sensation pricked my nose and throat . . . And I thought I would strangle. My gas-mask went on. I labored for breath. I sank to my knees, then struggled along again. "O.K. up here" came the voice, and I threw off my respirator. I drew freely of the fresh air, but my legs failed to hold me and I sat upon the ground . . . "Pretty rancid down here . . . Must be drifting around the mountain-point with the north wind."

We rested a few minutes upon the high jutting point over-looking the valley. Villers-sur-Preny lay below. Far to the west a silvery thread appeared around the hill and wound itself gently toward us, disappearing behind the Boche cemetery and coming again to view below the village. The spires in the town below reached imploringly to High Heaven to stop this THING . . . A pale yellow, smoky scum drifted across the low-lands clinging close to the surface. Through the broken smoke-currents we saw the flare of a burning town. To our fore across the silvery stream, white splotches, wraith-like apparitions, danced in and out of the fluid smoke,— the tell-tale markings of Germany's efforts to burrow into the lime-stone hill-side . . .

And I turned faint . . . I wanted water . . . With a none-too-gentle chuck in the ribs and an assisting jerk at my arm, Meeks aroused me from my introspection, and we trudged down into the village. I sat on the flag-stones before the German-kitchen, and my old top-sergeant brought me water. A whining shell forced us to cover in the basement of a stone-building across the street from the church . . .

"Seems to me to be a perfect place to spend the remainder of the day." He agreed and we lay upon the cold, stone floor listening to the constant rumble and churn of the long range guns, pummeling, battering, blasting at the remnants of our little French village . . . My throat was dry . . . My nose burned . . . "We'll wait awhile . . . It'll pass off . . . Just got your snoot full" . . .

We turned to talking of other days, of San Antonio . . . friends 'back there' . . . of his wife and happy days. We were weary of war and wanted to go home. One of the boys who had followed us below ventured the assertion that "this war is going to be over with me soon or I'll take a permanent French furlough." . . . I turned to Meeks to ask him if he recalled the statement of the German prisoner whom the officer had taken from us, when I asked him if he did not fear punishment at the hands of his American captors, and he had replied: "Was man nicht kann meide muss man willig leiden" (What one cannot avoid he must willingly endure). Then all of us wondered if the German's philosophy were not correct. One of our group with a philosophic turn of mind arose to point out a painting of the Madonna hanging askew upon the wall. A foot-print, clear impressions of hobnailed shoes across its delicate tints, showed it had one-time been upon the floor. It had been rescued but greatly soiled. Then our friend pointed to a carving in the soft stone of the wall, an attempt to reproduce the Madonna. In explanation, he ventured the assertion the picture was French; that the carving was German; and, although they fought and killed each other, they had a common source of religion. As he started toward the door to leave us he expressed the opinion that perhaps some of us would live to see the good people of the world gather for strength and unity around that 'Source of All Power',—the only hope for civilization!

A big shell screamed across the mountain top and disturbed us in our reveries. It howled into the valley on its flight of destruction. Instinctively we arose and peered out the doorway. The suppliant church-spire across the street splintered. A gaping hole marked the passage.[6]

We slipped into the street, waded the chilly stream while the water gurgled around our ankles, and forced weary steps up the hill-side. My breath was short and I stopped. A nausea pervaded me. My stomach revolted . . . I strained and retched . . . "Clop! . . . clop! . . . clop!" came unmistakable hoof-beats to my ears . . . The lieutenant of the sorrel horse dismounted. With an arm around me he asked: "Are you hit?" . . . "No, Sir, Lieutenant . . . Just smelled up a little too much of Jerry's perfume" . . . "Ride my horse . . . I'll walk beside you" . . . "I thank you, Lieutenant. I'm not going far . . . I'll be all right soon."

Morning came and seven men sat on the trunk of a tree which had been wrenched from its anchorage. A lowering sky threw a pall over a mizzly day. Raindrops clung to the brambly undergrowth. Someone produced a tin of 'iron-rations' and we ate . . . A putrescent stench arose and the food was put aside. "Hush up, Heine! Don't you see we are eating breakfast!" laughed one as he slid off the log to press back the bushes disclosing two maggot-infested gray-clad forms . . . Rusty rifles were lying near. . . . Steel helmets had sunk down over decimated boyish faces . . . And we walked away, disbanding, each taking a different direction to return to his company. Meeks hesitated and turning to me said: "Don't forget . . . San Antonio . . . when it's over.[7]

6 See hole in spire in picture facing page 209.

7 Sgt. D. E. Meeks died at Whittle Barracks, Prescott, Arizona, Feb. 29, 1920, from effects of gas received in the Argonne Forest.

With relief came relaxation. With relaxation came introspection, and I walked through the timber reflectingly. My orders were to report to battalion headquarters "today", but the day was young. Why hurry? Had I not seen the war-dog break his tether and, snarling, drive the fleeing fox before him? Had not the 'vulpes' flashed his yellow brush in our faces and scampered to the safety of his hole in the hill across the river? And now why should I hurry?

The churn of horses' feet in mud and water reanimated my retrogressive mind. I stood on the muddy brink of a scum covered pool. Horses, fetlock deep, half in mud, half in bubble-covered water, lunged to leave when I approached. A little frog hopped energetically, striking the water with a splash. He kicked mightily to return to the shore, and with submerged body, big limpid eyes protruding, he peered blinkingly up at me . . . "Must be fine, old pop-eyes, to have a bath when you want it" . . . and I shook my vermin infested clothes from my body and hung them on a limb. Wading into the water, with mud oozing between my toes, I dipped down and laved my body, while high above, a mere speck against a grayish cloud-beflecked sky, a German aviator flitted in and out, heading back toward the Hindenburg Line . . . I was having my first bath since going into the trenches!

Eight hours . . . four miles . . . and when the evening light gave evidence of its withdrawal from the sky for the day I drew up before the battalion P. C. A lethargic quiescence pervaded this sod-covered hummock from which protruded huge rough hewn timbers, supports for the roof over the heads of the 'brass-hats'.

"Coporal Emmett of the ammunition detail reporting to the colonel by order of Color-Sergeant Meeks" . . . "What company, Corporal?" . . . "L-359, Sir." . . . And many, many questions followed . . . "Any shells falling far back?" . . . "Are all the dead buried?" . . . "Are the roads clear?" . . . and I began to think the colonel knew nothing about the war.

A 'runner' came: he would direct me to my company . . . "Shake a leg before it gets dark . . . They will shell soon." . . . And we met Bill Goodson leading a dog, a German-police pup, a gaunt, sleazy looking thing. Bill pointed at the dog: "My buddy since I lost you . . . You look so much alike . . . We'll share him, now, that you have come back . . . Anything biting you? . . . You don't look so good . . . Take my bunk down that hole. I'll tell Big Tommy where you are. He'll understand".

I sank down upon a blanket-covered board. My back ached and I thought of the poor devil back in the cavern who had died that evening . . . "Yes, he had been gassed and had pneumonia, too."

A dozen men fidgeted about in the hole. I could hear them swearing softly . . . "Got it again . . . ---that madamoiselle anyway!" . . . Tread . . . tread . . . tread . . . the guard was changing . . . Then a long quiet, and the light showed through from above. I struggled to my feet . . . "Go back, lie down, be quiet, you sacrosanct fool . . . Don't you know you are sick . . . Big Tommy will be around to see you soon" . . . And then I knew that the oil of good-fellowship, running from soldier to soldier which lubricates the gears enmeshed to officialdom, had been caused to run by Bill's pouring.

Big Tommy, now 'top' since Meeks was promoted, came bringing with him hot coffee. We drank and talked . . . "You will not report to the company, yet . . . I'll tell Sterling . . . He won't peep."

Another day passed. Evening shadows laid long arms on the ground. A stooped figure, pack on back, disconsolately walked toward the rear. He did not pick the sheltered places as we went. The very droop of his shoulders showed he did not care. A soldier caught up with him and placed an arm around his shoulders . . . And he passed on . . . "It's old George Singleton . . . They've busted and transferred him . . . Damn nice whim of authority . . . He's been with us from the first."

L Company's new sector looked down from its concealment in the timber upon Pagny-on-the-Moselle. A gentle rise of the ground, crowned densely with trees and brambles, gave way to a rapidly descending plain of a quarter-mile expanse. We occupied the timber. Our men grumblingly, but ceaselessly, wielded picks and shovels. They were digging their homes again in the rocky soil of France. Trees clung closely to the river before us obliterating it from our view. At intervals through the winter-green leaves the white-chalk bluffs marked the abode of our enemy. And when an excited Boche pitched his voice above the guttural, it would be wafted over to us to vouch for his nearness.

A shell crashed out of the sky and broke before the company P. C. They carried 'Old Red',[8] the company mechanic, away. A trail of blood spotted the leaf-covered ground. Across his body lay the shattered fragment of his leg . . . Another shell whined out of the sky and pounded down the steps of the P. C. . . . And what a relief! It was a dud!

Lieutenant Nysewander ambled to the timber's edge and spied long and closely through his field-glasses. 'Rumor' soon had it that we would attack tonight. Rumor and counter-rumor ran through the ranks. When darkness came and after the men had taken to their holes a runner came with the order to "Roll packs . . . Be ready for the order to advance."

I had, now, thanks to Bill and Big Tommy, no assignment with the company. Then why should I go with the attackers? "But what's the difference", and rolling my pack in preparation, I sat down to await the order. Hour after hour passed, and nervous, fretting, cursing soldiers wondered why the delay: "Let's have it over!" . . . Then came: "Unroll packs and double the guard." . . . And perhaps it was Sterling who let it 'leak' that Nysewander's enthusiasm had been rebuked at H. Q. Those of us who heard it had visions again of the white flag on the walking stick the morning of September twelfth! What an officer!

My nausea had now left me. The pains in my back returned at less frequent intervals. Reporting to the P. C, I found I was assigned the position of non-commissioned gas-officer with duties carrying me throughout the sector. . . . "Observe what gas is falling and report." . . . What a snap? I roamed at large along the front, bedding down at night with Bill's dog when he was away on guard.

Returning one late afternoon through the timber and reaching a clearing cut through the forest, I started across, but heard the whine of a shell and beat a hasty retreat for cover. Two boys who were moving ahead of me made a break for the farther side. One boy fell. A blast rent the air. Cries of distress and anguish made me know someone was injured.

8 I have always had the impression that 'Red' Hurmans was killed by that shell, but the War Department records indicate the unfortunate man was Hurman's friend and close associate, C. P. Rankin.

Soldiers came from the nearby bushes to huddle around two prostrate forms. I ran across the clearing and looked. One poor fellow had fallen on his hand-grenades which he carried clipped to his belt. They had exploded and he had been severed entirely in twain. Ugly, bleeding wounds rendered his companion helpless.

From a position far out to our left came, late each evening, the staccato "chat" . . . "chat" . . . "chat" of a machine-gun. The bullets zipped and whined about our position. So persistent was the firing that our men began to dread the hour of his activity. An officer came into the shallow trenches (Trenches were always shallow when American boys had to dig them!) and commanded a sergeant to get a squad of men and "go get that machine-gun". The sergeant departed to remain away only a short time. When he returned the officer was waiting . . . "Did you get that gun, sergeant?" . . . "No, Sir: They were *using* it!"

I ranged along the hillside one day and saw soldiers concealing American rations in the bushes. I surmised we were to be relieved. I returned to our area with the waning of day. This time, thinking perhaps I was making my last trip of inspection, I skirted the timber far out toward the German position. From across the clearing in front of our lines came soldiers marching single file. It was too dark for me to distinguish their nationality at that distance. Fear gripped me. "Was it a German raiding party headed straight toward me?" Concealing myself behind a tree and deciding to fight it out then and there, I waited. They came steadily closer and closer. Now I could see American uniforms. An officer was leading them . . . and I recognized Lieutenant Frank Feuille.[9] I reached out, taking him by the collar. He fell back in fright . . . And we laughed . . . He explained he was taking a machine-gun detachment to position and had become lost. I gave him the directions and they went out into the darkness. A few minutes later I heard the "chat . . . chat . . . chat . . . chat" of a machine-gun. Months later while recovering in a hospital from wounds received in this skirmish, one of Frank's squad told me the story (not recognizing me) of

9 Frank Feuille, a University of Texas class mate of mine was last seen by me before this incident while we were en route to New York just before sailing for France. Frank came back to the United States after the war but died shortly thereafter. I never saw him again after that evening in the timber.

how a soldier had frightened his lieutenant just shortly before they had established contact with the enemy and he had been shot.

American officers concealing iron-rations in the bushes brought forth a confusion of rumors. "Another big drive", they prophesied; or "Relief"; or "Are we to stay here all winter?" By the time I had reached the company P. C. the "Early Morning Gazette" had it that Pershing had quarreled with the Allied Command; that he had been greatly disappointed when he had been called off from completing the infantry drive into Metz; that supplies were even now being stored under bushes in the rear for use in the re-attack; and that L Company was to be the initial point of attack. They had it on good authority, so they said, that we were to occupy Metz within a week.

A feeble sun receding behind a western hill threw flecks of light upon a thinly covered sky. The thrum of air-motors, now faint,—but swelling into more and more intensity,—brought us out from under cover. We craned our necks to see a passing squadron. Forty planes flying a wide 'V' formation passed swiftly over our position, Metz bound . . . And we thought of geese fleeing southward before the pinch of winter-blasts . . . Yes, they were geese, we all agreed!

Hardly had the monotonous thrum faded from our ears until the deep growling booms of dropping bombs told us of their arrival. The sky lit up in streaks. Shafts of light flitted here and there, sweeping the sky in search of the invaders . . . And when it was dark a faint drone passed over again. They were flying high. Their noise faded out in the southwest. They had returned home.

The following morning, at soldier's-stance, the opportunity came for dissemination of, and confounding, information, and the entire company was excited over the previous night-raid. Some had it for certain that "the last shot had been fired", others assured us it was "a farewell kiss before we leave". But days came and went. The hum-drum tedium of standing guard at night, of trench-digging . . . always trench digging, . . . of night-patrols to the very "hole-ups" of the Germans, caused onetime flippant and buoyant youths to falter in step and mutter stingingly sarcastic replies to their best 'buddy'.

But Bill's sense of humor never died. He caught 'outpost' guard one night when the shelling was most intense. He left me with his dog sleeping in a dug-out several hundred yards back of the line. Bill lay in the drizzly, cold rain and his mind went back to the days at Camp Travis when he thought "it could not get any worse", and he whispered to one of his buddies: "Keep an eye out . . . I'm going back for a few minutes" . . . And then the dog raised up from my feet and growled a soft warning . . . I heard someone calling me . . . I answered . . . It was Bill . . . and he just wanted to know "if I thought it had got any worse."

I moved into the company P. C,—a log covered cellar,—and slept in the corner with the "runners". Unusual activities began to show through the interchange of messages from P. C. to H. Q. Sterling's jovial face brightened and he whispered into my ear: "Won't be long now!" Lieutenant Nysewander went out into the night to hunt his dog . . . "O, that dog. Won't someone kill it?" . . . The Germans must have heard him calling for he returned reporting heavier shell fire . . . He must have had a narrow escape, as he became reflective . . . "Frank, your typewriter, please. (What a word to use in the army!) I want to write my will" . . . And he dictated: "Somewhere in France . . . Realizing the uncertainties of life . . . and being of sound and disposing mind (And I wondered?) . . . In the event of my death . . . "(Why did he not say, "Upon my death?" Could there yet be doubt in his mind?) . . . and we signed as witnesses.[10]

"The company will be without rations today" said a lieutenant. "The kitchen has gone back" . . . and my heart pounded with joy . . . "Sir, if the lieutenant would suggest: I saw the relief advance-party cache their supplies" . . . And when night enveloped us with its protective mantel, and from under a low crouching tree we distributed the last rations on the front. . . . "Thank you, Corporal: Good work . . . I do not know anything about this theft."

Crunching feet from our rear marked the coming of men. Dim, almost invisible forms in single file moved toward us. "L Company's relief,

10 Lt. Nysewander was promoted to captain and made commander of K company. He was killed along side of Joe Templeton. Some days afterwards, Company Clerk, Benj. Frank Sterling, I. Co., found the body of the Lieutenant and took from it, and delivered to the proper authority, the will he had written in the dug-out that night.

Sir", . . . and with a more sprightly step than ever before the remnant of the shattered company threaded backwards into the darkness. We were saying 'Farewell Forever' to the Saint Mihiel sector.

Chapter 11

Changing Fronts

The churning rumble of shells quickened our retreating steps. We were going away now. A pervading solace swept through our minds only to be replaced with an apprehension as each screaming shriek came nearer and nearer. Each booming blast, when out of harm's range, revived our spirits . . . And I recalled the teachings of the days of my youth: "I am in a great strait: Let me now fall into the hands of the Lord, for very great are His mercies, but let me not fall into the hands of man . . ."

"Quiet . . . Absolute quiet . . . No talking", . . . came the orders, whispered from man to man down the line . . . Then out of the darkness, pitched above the whistling whine, came the clarion call,—unmistakably the voice of Lieutenant Nysewander,—"Here puppy! Hi! . . . Hi! . . . Hi! . . . Come doggy" . . . (Then) "Damn that mangy hound. Won't some one please kill it?" . . . The dog ran along the line of men, sniffing, smelling, seeking his master . . . Nash unlimbered and fixed his bayonet to his rifle . . . There was a short quick jab and a cry of pain . . . Nysewander's dog was kicking convulsively at my feet.

Hour after hour the untrained guide led us here and there, through brambles, around bushes, until we came to the crest of a hill above a valley. We were far to the right of the descending road. Sliding and slipping, the indistinct file lowered into the valley. Shifting clouds granted a watchful moon a fleeting glimpse of staggering men, worn by war, weary of foot, as they relaxed in muscle by the relief from the mental strain. Wading through the cold, blue-black stream, motionless to us at night, the column strained up . . . and up . . . until finally the timber-crowned hill beyond was gained. "Death's Valley" had been passed.

Sterling stumbled . . . He regained his feet but went down again . . . "I'm all in, boys; go on; don't bother about me." . . . "Not yet, Big Boy! Not by a heluva sight! . . . We're nearly out! . . . Don't you understand? We're nearly out!" Big Tommy and I took his pack and rifle, and with coercing arms around him,—passing, forever passing prostrate anguish-ridden forms upon the ground,— we trailed the company's march. A cross-road . . . and the order came back: "Fall out." The men sank down in the water-soaked ditches like cattle hard driven on the trail.

My next conscious moment came when Old Sterling was yanking at my arm. I sat up. Day had come. A hoary frost shimmered white as a feeble sun strained to light the return of a new day. My arm, benumbed from exposure to the frost, failed to respond to my efforts to raise it. Sterling gingerly placed a filled mess-kit on the ground before me, and with gentle strokes upon my deadened arm, grinned the comfort: "I saved breakfast for you. Better eat now. We're going again."

The crumpled heap of Fey-en-Haye now lay behind us. The timber of our 'jumping off place' of weeks before was browning by the invasion of winter's blasts. Swinging packs again upon stiffened backs, and bearing to the right, skirting along the shell-thumped rim of a hill, we passed before a "hole-up" from which protruded the kinky head of an American negro . . . "O, Lordy! . . . Lordy! They done let some ob you come back? You ain't been killed, is you?"

From our rear now came only the faintest rumble off the front. Buoyant with the contemplations of the immediate future, our pain-racked feet swung rhythmically across the broken sod. Jovial banter again passed our lips. Faces, long furrowed with distress and strain, were wreathed in smiles. Feebly one voice . . . then a chorus . . . sounded louder and louder as the rhythm marked the cadence of scuffing feet:

"'T was a long, long trail that wound us
Into No-man's Land in France,
Where shrapnel shells were bursting,
Disputing our advance.

There'll be lots of drills and hiking,
Before our dreams yet come true,
But we've showed the yellow Kaiser,
What the doughboys really can do."

From out the distant timber came the throb of motors. Trucks bumped over a rock-strewn road. Once we gained the course, a peacefulness pervaded us even though constantly being crowded aside with the cry; "Give 'way to the right."

Pup-tents were spread here and there under the browning green canopy. A man stood up suddenly, and shouting, ran toward me: "Chris . . . Chris . . . With those francs I bought you something to eat . . . I knew you'd come . . . some day." It was wobbly, pudgy Joe Templeton. He ran along by my side, carrying my pack, now my gun . . . talking. "Hand is all right now . . . I'm catching my company when it passes. Will see you tonight."

We gained the top of a hill on which lay the litter, the remnants, the left-behinds, of many a passing battalion. It looked like the grave-yard of discarded materials of war. Ahead, the terrain sloped gently toward a placid stream on the very brink of which nestled a peaceful French village. Directly beyond and towering over the valley arose a mountain. It was crowned with low squat stone-buildings. The valley above the bridge, which marked the entry to the valley-town, widened out fan-like. The deep green winter-verdure marked it as a refuge for peaceful solitude. Lieutenant Baker walked by my side. He pointed: "I feel like lying down there . . . forever." . . . "And when you awaken, Sir, I will be lying by your side . . . with a bottle of 'schnapps.'"

A lazy sausage-balloon tugged gently at its tether. The basket hanging below its belly had two heads showing over the side. These men were waving at us. A cloud floated almost motionless above the big bag. As we looked a German plane shot straight down from out the cloud and was upon the observers in a moment. There was a streak of flame. The two men dived from the basket. The balloon burst into flames while the opening parachutes gently lowered two American soldiers to the ground.

We went on down the hill talking about the balloonists . . . It was not safe even this far back. We came to the town. It was Griescourt. We had come to our first stop behind the lines.

We located the company P. C. in the upper story of a stone buildings. Argie, Big Tommy, Sterling and I lay down for a rest. Much of the evening had gone before we went upon the streets. Griescourt was an "Off Limits" village, so we got a G. I. can and climbed the hill to the city above. An officer walked hastily before an "Off Limits" sign. I recognized him as Wright Felt, a boyhood chum, but we were bent on other things and he was in a hurry.

The descent from the heights was negotiated with the utmost care, for we carried the beer for tonight's celebration. Sure, we were going to celebrate! Had we not gone over the top and come back safely again?

Hours later the remnants of the 'old gang,' with high pitched voices, were trying in inharmonious concert to tell of the days just passed. One now deep in his cups was paraphrasing Burns' "A Toast":—

"Instead of a song, boys, I'll give you a toast,—
Here's to the memory of those of the 90th we lost.
That we lost, did I say? Nay, by Heaven, that we found,
For their fame, it shall last while the world goes 'round."

And just at that moment the door opened. A stern faced lieutenant appeared. With raised cups he was greeted: "Just in time, Lieutenant. Have one!" . . . He grinned: "That's my trouble now, boys. Thank you."

I drew the assignment of non-commissioned salvage officer the following day. There was no falling gas to report so I arranged for the collection of the discarded clothing of the men. It was a busy day. The necessary exchange of clothing and equipment had scarcely been completed until the big hill had spread out its evening shadow to cover our village. The bugle blew for us to "Fall in" again. We realized we were in for another night march. It was safer, of course, to march in the night time, but men no longer cared for safety.[1]

1 (Ed.) "Salvage piles" and "salvage depots" were areas of discarded items of clothing, shoes, weapons, and other equipment collected on a battlefield to reissue to other troops in need or to send to the rear for repair if necessary.

King sat upon the ground nursing a boil on his neck. The captain appeared stepping airily. Our commander was an unusually large man through the hips, and his ill-fitting jacket,—probably commandeered in an emergency while on the front,—stuck out in flounces around the hip-line, giving much the appearance of a peplum or ripple-tail blouse. The flounces tweaked upward and receded with every movement of his legs. He stopped before King: "Any trouble, King?" The explanation followed. And with a grimace the captain replied: "A ride tonight in that truck . . . with a bottle of cognac . . . don't you think . . . would turn the trick?" And as the company marched away in a drenching rain, King's head appeared above concealing blankets. He waved a bottle!

Throughout the night we trudged along. Roads were now clear of traffic, and the resounding whacks of hobnailed shoes clattered and crunched toward an unknown destination. The eastern lights began to show, and we lay down in a deserted village. The floors were clean. What a wonderful relief! Breakfast was served, and again we were on the march. We headed down a valley across which struck a clean white road intensified in its whiteness by the purple of abutting vineyards. Ranks were broken by the men at every turn. Scowling officers reinforced the protestations of the horrified Frenchmen as we skipped from the road and bore back in laden arms, grapes . . . luscious grapes . . . dripping with juice.

From off the distant hill came a caravan of trucks, rolling head to tail, interminably, across the plain. All day long they moved as one immeasurable snake, brown-coated, crawling westward . . . and where? Our paths intersected and we abided our time to pass between. Stoical Japs, immobile in their driver's seats, looked neither to the right nor left as they sped past us. Ahead of us there was familiar topography, a menacing hill, and a small village nestling against its breast. We recognized our old home . . . Pagny! How glad we were to return to this familiar scene!

I lay in the hay loft above the fragrant pens of a frugal Frenchman debating to myself. Should I go first for a bath, or would not 'euffs' and vin-blanc be more delectable? I could bathe by myself but to eat euffs and drink French wine without Joe, King, or Bill would be unthinkable! Then someone called my name. A new recruit, climbing up the ladder to our hay-loft, stopped with amazement when my name was called and inquired: "Are you Chris Emmett?" . . . "Guilty again . . . What's the offense this time?" And he told me: Back at the hospital he had been very sick. A wounded boy, fresh from the front, was next to his bed. He called continually: "Chris! . . . Chris! Chris Emmett . . . I wonder if he made it through". Since the recruit had never heard of me before and did not know the name of the wounded lad, neither of us ever had a clue as to his identity.

I decided that ablutions might wait, but would not Madam be pleased, again, to serve a dust-covered bottle and listen attentively to experiences since our departure? Would not King soon be needing someone to "split a bottle for the sake of the boil on his neck?" I scrambled down the ladder, looking carefully and stepping gingerly lest the old Frenchman had only recently incarcerated his cow,—and rollicky, rolling laughter directed me to the "place on the corner". Backed up against the window with his feet protruding over the top of the table, King was the center of attention . . . Bottles popped . . . "Make it another" . . . and the Battle of Vin Blanc started all over again!

An artilleryman, sporting a close cropped mustache, raised a bottle above his head with gusto, and waving his other hand to one wearing a similar insignia, shouted: "Here's to the boys of the 'Big Fire Works'!" A doughboy with steel-helmet askew, his face bristling with a whisker-growth of many days (so long had it grown that it turned back slightly at the ends), slipped his helmet-strap from under his chin, and in clumsy movements fanned his face with the 'tin-lid'; "Plenty of hot air today." . . . Then turning to the convivialist with the tonsorially perfect mustache, he remarked dryly: "Here's to the mud on your face!" The rousing guffaw from the infantrymen drowned the retort of the artillerymen, and he subsided to pour again from his bottle while his pink cheeks reddened from the insult.

A long, lean, lank figure, with broad shoulders, tapering down to the very feet which spread expansively upon the cobbled-pavement, loomed before the window. It was Major Collins. He hesitated only perceptibly to peer in. Then he dipped his head slightly forward and was gone. Someone, sensing that he did not wish to disturb the men in their relaxation, started up the improvision:

"O, he's a jolly good fellow;
He's a joly good fellow.
There're some we know who were yellow,
But of him it was all a d --- lie."

The room reeked with the fumes of the 'makins', and Monsieur . . . with his drooping mustache and his box-cut beard,—who had been too old even when the war started to go to the army and who now stayed at home to ingratiate himself with the passing Americans,—standing on a high box to take a bottle from the uppermost shelf, in obedience to the specific demands of one who waved large franc notes at him, . . . appeared as a headless apparition as the smoke settled densely against the ceiling. . . . King laughed when I pointed at the Frenchman on the box and suggested I was 'seeing things'! We would have poured another bottle, but the scurrying clatter of hobnails on the flag-stones suggested impending events. The bugle sounded: "Fall in". I shambled uncertainly to my feet to find that only King and I remained. King sank back to his comfort with: "O, what's the use. I have a convenient boil on my neck!"

The company was drawn up in the street when I appeared to slip into my position. An understanding buddy to my rear whispered: "I'll watch you", and wavering, weaving with every note of 'Retreat', my knees sagged at the echo. King's timely appearance resulted in our departure, at the close of the formation, toward the bath, now long delayed. We found a shed-arrangement one side of which was devoted to the use of the 'men' and the other to 'officers'. From out the door under the marking: "For Officers Only," came Lieutenant James A. Baker, Jr. King and I, arm-in-arm, led the parade. Unlocking the door of "For Officers Only" and shoving a towel at us, he said: "In there you 'soaks' . . . and soak good." Presumably he knew that a 'man' divested of his outward habiliments would not be mistaken for an officer.

We showered hot and showered cold until the hurly-burly world no longer reeled around us. When we returned the towel, he grinned the assertion: "You look better and I hope you feel the same. Now go to sleep."

After a short nap I came down onto the street to stand in the cool, crisp, clean night-air. Many things had happened since I was here some weeks ago. I looked against the mountain-side and recalled lying up there watching my first night air-attack. I could not but wonder when would this terrible thing be over. The faint booms from the now distant front broke softly upon my ears. These sounds had a much more significant meaning to me now. Then I heard the rumble of a motor in the air. Shafts of light darted skyward from the hillside as it had done months ago. Then I had thought it a grand sight. The light-shafts were switching again here and there in search of another invader. The same scene over again! Then Bill came tearing down the street colliding with me: "What's your hurry?" . . . "H , I've learned something since I was here last!" He disappeared in the darkness headed for a cellar.

A clear morning came. "Light packs" were ordered and we scrambled up the trail leading to the mountain-top. "The colonel will inspect in maneuver for possible recommendations for promotion in rank" . . .

Captain Whitaker placed me in charge of the company, and with a "Squads Right! Forward March!", and stepping the cadence firmly, I watched from the corner of my eye. The captain diverted his attention to the colonel. I zig-zagged . . . again . . . and again. He looked back from the colonel with his face wreathed in a smile. Then it froze. "Company halt! --- --- ---:trying to make a snake out of the company! . . . To the rear. Fall in, and stay there!" . . . Then I knew I would be able to stay with the boys.

Ross' wound at the front had detained him only a short time at the hospital. Since we did not drill in the afternoon, Ross and I, with some others, decided we might spend the evening with some degree of pleasurable variation if we could inspect the fort built into the top of the mountain. We, therefore, appeared before the entrance but were greeted with the firm refusal of the French in command to permit any Americans beyond its portals.

We found the fort to be a huge stone structure built below the level of the surface of the mountain upon its most extreme height. The fort was capable of quartering thousands of men. Around the building ran a moat which effectually prevented approach from the land, except at the entrance of the railroad, the tracks of which sloped down to a steel-enclosed entrance. From the level of the mountain,—as the building was cut deep into the rocky mountain,—it was most difficult to see that an impregnable fortress was there.

Arguments in all available languages brought from these Frenchmen only a repetition of their refusal to permit us to enter. Then we decided to 'scheme' our way in.

Ross and I scrambled along the moat-side, and clung to the rough, protruding stones until we came to a jut in the wall. These rough stones in the angle of the wall permitted insecure footing from the moat to the top of the fort some thirty feet above. Ross and I had been selected for the task because in size we were the smallest of the 'visiting delegation'.

I and L Companies going into the Argonne. Arrow points to author.

We climbed inch by inch, each assisting the other,— pushing and pulling! We were finally at the top, but to our dismay the stone rampart was crowned with long, pointed steel-barbs, the tips of which raised above our heads.

But we must go over! Ross raised me as high as he could, and I crawled upward, spreading my weight carefully as I sprawled upon the tops of the tines. Their points dug into my body. Working gingerly I released myself and dropped to the building top . . . on the inside! To bring Ross over was an easier task. This having been accomplished, we sought an avenue for descent into the fort. We tried to slide down the muzzle of a gun, but my hips stuck and Ross had to yank me out by the feet. We tried a sky-light, but there was no rope by which we might lower ourselves. Then we pried the facing from a door and stepped below the roof. Beneath us, neatly fashioned, were the bunks for many thousands of men. They were little used now, for the war had moved away to the north and to the east, nearer to Metz, where men dug into the cold clammy clay and denned up like wild animals. O, how those same men, fighting those few kilometers away, would have been willing to give their right arms for a chance to bunk again in this clean fortress.

We reached the inside before the doorway, where the railroad enters the structure, amid a clamor of French protestations. They shook their fists and ranted volubly. We explained the necessity for "your good American allies seeing your wonderful fortifications". But they were not to be mollified . . . Then, "Throw us out if you like! . . . We are here . . . Whatcha going to do about it?" . . . The Frenchmen grabbed us ,and the gates swung open . . . A score of Americans threw back vociferating Frenchmen . . . and we were on the inside . . . all of us!

A French officer, commanding a sprinkling of English and possessed of a sense of humor, suggested "a drink to the gallant American soldats . . . and perhaps they will then depart and not again violate the commands of their brothers-in-arms." . . . We did . . . both!

We arrived upon the streets of Pagny again in time to learn that orders had been posted specifying that certain non-commissioned officers,—

whose records of conduct, while in the 'rest area', permitted,—might upon applying for passes be granted a leave to Toul. I had no reason, however, to believe that my name would be on the list, since, no doubt, my conduct before the colonel had caused a moiety of chagrin to a certain captain of my acquaintance. I started, therefore, toward the company P. C. to find out what I could.

My progress was momentarily interrupted by the appearance of the major, who stood menacingly although grinning, before King who was displaying a shining 'iron-cross' which he claimed he had taken from a German prisoner at the front. The following conversation occurred:

> Major Collins: "Give me that iron-cross, King. Don't you know that all such things taken from prisoners of war must be dispatched immediately to headquarters?"
>
> King: Sir, I so understand the order . . . but . . . but . . . if you want an iron-cross why don't you go capture yourself a Dutchman?"

With a guffaw the major saluted and walked away . . . empty-handed!

The captain's passes to Toul were being issued by Argie. My name was not on the list, but the captain listened to the argument and said: "Just as well let him have one. He'll go anyway."

Off to Toul we trudged. Bill led the way. Our backs were turned to a cold, misty wind which, with the coming of the evening, swept off the mountain to send shivers through the bodies of men much weakened by the trials of the weeks gone by. Toul, a walled city, lay some six kilometers to the southeast. It was a city of no mean proportions. But it was now 'Off Limits' to all soldiers except to those who held passes to report for duty behind the walls. We learned this when we displayed our 'company-commander's pass' at the city's portal.

Adamantine guards, accustomed to every form of attempted evasion to effect entrance to the city, and fearing nothing except detection of violation of orders, refused our passes . . . "The line-captains will never learn . . . We must turn you back". A captain of the M. Ps. approached . . . "Now, what you soldiers want?" . . . Sarcastically one of our party sneered: "Sir! Balaam may ride his martial transports in triumph up in the front, but in rest-camps his master's keepers with-hold the wisp of hay." . . . The M. P. captain half turned his back upon phlegmatic sentinels, and in sotto voice asked: "Just back from the front?" . . . "Sorry, boys, but it's orders" . . . He walked a few paces away from the guard, and concealing his half-raised left hand before his body, jerked his thumb in a pointing motion toward the left: "Half a mile east, there's a log over the moat . . . and Good Luck!"

A gleam of understanding lit our weary faces. We took up the march again toward the east, keeping close to the water's edge in search of the log. A high stone wall obscured our view of the city. Only here and there slate-colored roofs rose above the ramparts. Turret after turret marked our progress, but we observed that they were now un-manned. They were relics of the past. These seeds, a menacing threat to neighbors, had lain in fallow soil. But the leaven had again activated. An ill wind had blown out of Serbia. The elements had struggled. The storm roared and we, infinitesimally small factors . . . insignificant in importance, except to ourselves . . . stood before this, one of the incubators of the world's woes.

The shadows of the evening grew long. Our progress was now toward the south of the city with its forbidding-moat and its unscalable wall. A woodland encroached upon the moat, and we found a trail marking the passing of other feet. Across the darkening water, an end mired down in the mud, lay a log slimed over by the sliding of other feet. We had arrived!

The foreboding wall beyond gave no clue to the entrance to the city. Less carefree spirits might have viewed the forbidding height beyond the log's end and turned heel to trudge disconsolately homeward—to lofts of pungent hay—but Bill's high pitched voice, halted by paroxysms of coughing, inquired: "Well, we said we'd go to Toul, didn't we?" . . .

With long legs dragging, bent at the knees to keep his feet from the water below, he sat upon the log, and propelling himself 'coon-fashion', disappeared into the shadows against the wall . . . In this manner did all of us arrive at the angle in the wall below the turret. Human- ladders, feet on shoulders, provided the means of mounting the top where the city stood before us.

The dim outlines of a rickety, shingle-covered building extended halfway up the city-walls toward us. We surveyed the precariousness of the jump. With an ironical 'suave qui peut' (Save himself who can!), I took the leap into the city below. Shingles rattled from the roof-top, and I jumped again. My feet tingled from abrading stones . . . and gingerly we limped into the streets of the city of Toul.

There is joy in anticipation, disappointment in realization! Streets, just streets filled with the noises of crunching feet, and lights dimmed against the expectation of air-raids, greeted us coldly. For what had Ave come? Now, what could we do? Vin blanc was nothing. Had we not slumped low from yesterday's invigoration. Certainly, there was an 'Off Limits' sign emblazoning the path and acting as an enticement to those who were oblivious to its dangers. Why should we go there to assist the lechers of France in their supineness!

A seductive odor permeated the air. It decided our problem for us. We would eat. We entered a dimly lit house and sat at tables. A smoke-filled room warmed our weary bodies. We would have 'steaks and German-fried potatoes.' . . . "No? . . . You will not fry them 'German-style'?" . . . Well, we don't blame you Frogs . . . We don't think much of German-fried potatoes, either . . . Just fry them anyway . . . just so you fry 'em."

A shrill whistle blew in the street. A voluble 'caqueterie' old Frenchman, hissing his fright, waved his arms to us in warning: "The gaucherie gendarme! The M. Ps. . . . They make the raid . . . They throw you in the brig!" With the remnants of our steaks in our hands we slipped out into the darkness, easily evading the truncheon-wielding cordon . . . "O, hell, What's the use? An M. P. to arrest you every time you sit down. Let's go home!"

At the gate the snarling sentinel, amazed at our reappearance,—this time from the inside of the city,—"swore and bedamned that if he couldn't keep us out, then he could throw us out" . . . but the captain smiled wryly as we left the city to trudge off toward Pagny.

The moist chill wind sweeping off the mountain held steadily in our faces as we marched off toward our billets. Bill said he would rest awhile. Perhaps his 'coughing would subside.' ... and we left him by the roadside. The chill-wind carried a mizzly dampness. When we reached the confines of the town a sentinel challenged us: "Halt! Who goes there." ... "American soldiers returning from Toul" ... "Welcome to our mist, buddies!"

Morning came and I inquired at Bill's bunk ... "No, he has not come in ... An ambulance picked him up ... He is now in the hospital in Toul ... with pneumonia." ... When I heard this I was frightened at first, and I thought of dear Old Molly, and Clarence, and Tip ... and I was lonesome for him ... and afraid that cough he had yesterday was worse than we had thought ... Then I remembered there was another kind of a world than this one we were living in; that 'it couldn't get any worse' than what Bill had been through; and perhaps he was in that other world of clean sheets and pajamas, and good food and plenty of it, with a kind, gentle woman all dressed in white, who sat on his bed while she coaxed him to eat. And I concluded if Bill were in that world then, knowing him as I did, he could fight that battle all right, so I went off to find Joe.

The papers printed in Paris[2] had come up that day and Joe brought one over to my billet. "Quentin Roosevelt is dead." ... "Poor Teddie ... but there are many Teddies, now" ... "And Wilson would not let him come over with a division. 'No amateur's war!' The hell it isn't! What do any of us know even now?" ... "Another Liberty Loan Drive going over in America ... And Pershing had promised us "Hell, Heaven, or Hoboken before Christmas!" ... Haven't we had enough of the first? ... Me for Number Three, and I'll find heaven in Texas unassisted."

Another soldier, hearing the trend of the conversation which savored, unconsciously, of the poignancy of our loss of Bill and the unquestioned desire to go home, broke into the conversation with: "What 've you got to complain about? Listen to this". And he read:

2 The Stars and Stripes, and The New York Times.

"You were right, Mister William T. Sherman,
When you uttered that message divine,
For only today have I laid them away,—
Those two buddies of mine.

We had crossed in the steerage from Texas,
..To the land that the papers call France,
We had buddied together in all weathers,
Together we'd taken our chance

And many a time in our hiking
When I was unable to crawl,
They carried the pack that was slung on my back—
With never a kick at it all.

Rugged they were, tough and sturdy,
Though maybe they never would shine
In a highbrow cafe on the rue de la Place,
But they were genuine buddies of mine.

And now that their ditties are finished,
The thought that is left to console
Is: They were rough, but made of good stuff,
And each of them harbored a sole.

So thus, when the snow fell this morning,
And keen as a whip was the air.
My buddies checked in—I was stripped to the skin,
The sergeant then issued a new pair!3

3 This was read from an issue of THE STARS AND STRIPES, the issue and author of which I do not recall. The Stars and Stripes, a soldier-published paper was very popular in France, keeping the spirit of the men very high.

Joe and I threw down our paper and joined in the laugh. But from across the street, where other lonesome doughboys were having a rendezvous in their hay-filled loft, came the chant:

"I wanta go home; I want a go home;
The bullets they whistled; the cannon they roared;
I don't wanta go to the trenches no more
O, ship me over the sea
Where the Allemand can't get at me."

Reminiscence and by-play could not go on for long. The familiar bugle call, followed by the shouts of sergeants: "Outside! . . . Full packs",— came to us, and we were soon in the streets again. With crunching feet company after company marched rhythmically . . . tramp . . . tramp . . . tramp . . . to the village-edge. There stood trucks . . . a sea of trucks, head to tail . . . Slant-eyes peered from under their canopies . . . Our stoical Japanese friends had come to take us for a ride . . .

"Where do we go from here, boys,
Where do we go from here!"

Rumbling out in an unbroken line, the trucks filled with standing men, too full to permit sitting, laid our course toward the evening sun. Town after town faded behind us. No longer were there evidences of shell-smashed houses. The vineyards,—untrammeled, dark green glinting splotches of silver and purple,—selvedged a contrast upon our white winding road. Old women, shambling on wooden shoes, carried their baskets piled high with the products of their vintage.

We ran for a long time westward, but no one would tell us our destination. Then we turned northward and knew we must be going in the direction of the Argonne Forest. My feet burned. Someone was standing on them . . . I cursed and he moved over onto the feet of someone else. When he, too, complained, we took up the chant:

"Iodine will make you happy,
Iodine will make you well . . .

but the fellow next to me glowered his complaint with: "If you step on me again you'll never need a C. C. pill!" . . . Then all of us laughed and wondered for the eleven hundredth time where we were going . . . "It was better at the front than in this thing . . . We could sit down there." . . . Some one in the truck ahead started singing. We joined in the chorus:

"O, the infantree, the infantree,
With dirt behind our ears,
The cavalree . . . artilleree . . .
The --- --- engineers,
But they'll never rest the infantree,
In a hundred thousand years!"

A raucous horn shrilled to our rear. "Slant-eyes" stuck his head from under the canopy and looked back. He swerved the truck violently to the right and we fell in a huddle against the side of the car just as a general-officer's car swished past us. Then a doughboy's voice raised itself above the hubdub:

"If you want to find the generals,
I know where they are!
If you'll want to find the generals,
I can tell you where they'll be.
I saw them!
I saw them!

D
o
w
n

i
n

t
h
e

d
e
e
p

d
u
g
-
o
u
t
!”[4]

Night came and it was cold and bitter. Our feet ached and it was a comfort when someone stepped on them and made the blood run again. Finally, the truck stopped, but it was too dark to see what the trouble was and no one would tell us. We stood on one foot, then the other, and tried to jump up and down to warm our feet, but the truck was too crowded.

4 (Ed.) “Slant eyes” referred to the truck drivers the French brought to serve in France from their colony in French Indochina that includes the modern nations of Vietnam, Laos, and Cambodia. In addition, a large number of Chinese nationals worked as laborers in France during the war.

We swore at 'Slant-eyes' and "knew well he didn't know the road" . . . Then an order came back to us,—a call from truck to truck in the darkness,—to "spill out on the side of the road, the right side" . . . After that we stood in the mud. It came over our shoe-tops, and the cold slime on top of the mud oozed through between the tops of our shoes and our 'wraps' . . . Pains coursed up and down our legs like electric currents . . . and our backs felt like they would break under the heavy packs . . . We shifted our packs and stamped our feet, but the glutinous slime sloshed onto the man next to us and he "wished to hell you'd stand still" . . . We threw our packs in the mud, and wriggled our shoulders. We felt like a huge burden had been taken away, but an officer came along and made us pick them up again, and the cold mud from off the ends of the packs now soaked through our jackets, and shivers ran down our backs . . . The officer took our names and numbers and swore he would have us court-martialed . . . but in the darkness no one gave him the correct name or number.

We could see through the faint light that men were working with a truck stuck in the mud ahead of us. We wondered why someone did not have sense enough to let the whole company of men get a hold on it and shove it out . . . But nobody ever thinks in the army!

A dark hill to the west gave us a solemn peaceful feeling, now, after we had been glimpsing the eery figures moving mechanically around the truck ahead of us. The blackening gloom of the hillside gave us a semblance of mental comfort, but the rigors of fatigue and cold were yet upon our bodies. Ryan of the medical detachment heavy of body and sore of foot, gingerly raised a foot from the freezing slush, and plaintively in his misery knew "it will soon be over for all of us. We can't stand much more of this."

An order came back, soldier to soldier, and we started moving. The mud had turned frosty and it cracked under foot. But it was 'something' to be moving again. We marched out of the slimy valley toward the hillside where we found barracks with dirt floors. We lay down on crude beds covered over with wide-mesh wire. They were built up high, one bunk above another, like book-cases . . . But we were out of the cold air now, and our feet once more on firm ground tingled and got warm. Day came again before hungry men left the barrack in search of food . . . There had been no bugle-call that morning, and those of us who had been to the front before knew what that meant. Why let Heine know where we are by blowing a horn!

Officers made a methodically careful survey of the ammunition now in the possession of each man . . . Perhaps we had thrown away some of our burdens!

Automatic pistols were issued, but there was no effort made to show the men how to operate them . . . It was taken for granted that all men knew this. What a crime against untrained youth!

I was made mail-censor for the company and told to read the letters as fast as I could "for there would be many letters today", and then "this may be our last chance to get the mail back to the rear." I read hundreds of letters as the boys put them on the table, and let everything go through I thought would console a father or mother back home.

Written in a neat script I found one:—

"Somewhere in France,

(And I clipped out the date.)

Dearest Mother:—

I am now at (clipped again), and I cannot but think what a short time it has been since I was with you and since I joined the army. I have today been assigned to this company, but so far have had no training as a soldier. I think we are going into a battle. The other men who have been in battles before think we are also. I have not been taught, yet, how to handle my rifle but they gave me one sometime ago, and I have been carrying it around with me. Today, they gave me a pistol, but I have never loaded nor fired one in my life . . ."

I laid this letter on the corner of the table away from the others and weighed it down with my pistol. I read on, one after another, mechanically, without understanding. My thoughts were hovering over that message for a mother! Perhaps it was a warning to a nation, too! And the pistol looked ominous as it lay there . . . "Unprepared? Was this the tauted army invincible? Was this the result of a national policy, the work of long-haired, fluent, peace-at-any-price men? 'I didn't raise my son to be a soldier.'"

I sealed the envelope, initialed it properly and slipped it into the home-bound sack with "Perhaps he may not serve well here, but surely there is a lesson there. It may serve at home."

With night came Sergeant J. O. Allen walking softly. In subdued tones he called me: "Come" . . . And we threaded back into the darkness . . . He was being given an opportunity to return tonight to an officer's training school . . . "But what will I do without you boys?" . . . I gave him my hand: "Go, J. O. . . . tonight. It's just a suspension of sentence . . . If this damnable thing continues through the winter none of us will be here anyway . . . And 'Good Luck'!"

Long before the murky canopy had been pushed back from the skies our tramping feet were striding westward again. Now there were no 'Slant-eyes' with trucks to haul us, and we complained bitterly that 'everybody rides but the infantry'.

We came to a French village which had been abandoned by the population because of its nearness to the front. As we swung our long column into the main street—a French village never has but one main-street—we saw a cinema cameraman on top of a building grinding his machine at us. The boys up front waved their hats at him and shouted good naturedly: "Under cover. He'll shoot you!" Lieutenant Baker, 'Big Tommy', Sterling and I were near the rear of our company. I shoved 'Tommy' out of line and we held up our hats to mark the picture for future identification.[5]

On and on we went. The distant hills showed splotches of rusty brown. We had seen the earth torn up that way before. Soldiers no longer rode in trucks. Mule-drawn wagons clucked along stony roads miring deep in the mud. What trucks we now could see were filled with food and ammunition. These vehicles moved faster than we could and we gave away to the right and scrambled out into the oozy ditches again, and when we got back into the road we stamped off the brown sticky mud from our heavy feet. The

5 See picture opposite page 224 with arrow pointing toward Big Tommy walking out of line. I am immediately beside him on the inside.

eye could carry only from hill to hill but there was no end to the drab centipede with its in- numerable legs working irregularly forward.

"I never knew
What the war's about;
But it looks, by gosh,
I'll soon find out.
So, my dear,
Don't you fear,
I'll bring you a king
For a souvenir;
And I'll get a Turk,
Bill Kaiser, too,
And that's about all
One feller can do."

The men up front had quit singing. "Chug . . . chug . . . chug . . ." came the sound of a motor sidecar . . . Now the long line of men refused to give away to the right to let it pass. The motor slithered into the ditch . . . Two M. Ps. got out and stood in the mud and tried to push it back into the road again . . . "Give us a hand, buddies!" . . . "O, yeah? . . . Ha! Ha! . . . Who won the war? . . . The M. Ps."! . . . Echoing down, the line went the call: "The M. Ps. . . . Some !" They had to get their own car out, for no one would help them!

Eyes now were fixed ahead upon a maddening, distorted devastation,—a valley from which had been swept every vestige of growth. Holes, multiplied one upon the other, told of the carnage which had gone before us. Vultures swung lazily in the air only tilting a wing now and then to maintain a gently sliding balance. They were gorged with human carrion but were spying out for the days of their leanness. But such days would not come so long as men slaughtered in frenzy!

Stumps, up-rooted timber, fragments of trees once stately and tall, lay upon the shell-slashed, pulverized crust of man's habitat . . . And some one sang, for it would not do for soldiers to let the fancy peer into the recent past lest he also look into the future:

"O, Madamoiselle from Armentieres,
Par—le—voo!
O, Madamoiselle from Armentieres,
Par—le—voo!
Hasn't been kissed for forty years,
Hinky . . . Dinky . . . Par—le—voo!

When I looked at that devastation and knew what that long line of boys must be thinking and saw how well they concealed their thoughts even from themselves, I could not but think that Lieutenant Dazey's Tribute was written especially for the boys of the 90th[6]:

"There is a tumultous confusion a comin' down the road,
An' the camouflage don't near-ways hide the dust,
An' it ain't no flock of lorries, though some's bring up a load
(I guess the provos winked—or got it first).
But now it's comin' closer; you can tell 'em by the roar;
It's the Umpty Second infantry a goin' in once more.
O, they've met the Hun at the length of the gun,
They've seen what he is, and know what he's done,
So that's why they sing as they slog to more fun."

We came to a zig-zag in the road. The trucks were having trouble passing over, and we sat down in the mud to rest. When a truck would mire down, its wheels would spin and throw a red slime over us and the driver would be admonished to "hold her, Big Boy, and watch her kick". . . . Then everybody would get up and shove.

The break in the road had been caused by an explosion, perhaps an air-bomb. The explosive had landed squarely in the road. Muck had been

6 Tribute, also, may be found in an early issue of Stars and Stripes. It was written by Lt. Dazey.

thrown far out into the adjacent timber, leaving a hole twenty-two feet deep by twenty-four feet across its narrowest irregular diameter.[7] We passed around this terrible rent in the earth and trudged along again.

It began to rain on us, but we went as far as we could before we sat down again to take a 'little leg ease'. By the side of the road was a lone soldier sitting on a steel-helmet for a 'swivel-chair'. He had the muzzle of his rifle turned up so the water could run down the barrel. He said he was the guard on the salvage-dump, that he had been there for days, but as long as he stayed there he would not have to fight and there wasn't 'anybody who could trade him out of his job'.

The salvage-dump was immediately behind him. There was a high stack, a veritable young mountain, of discarded uniforms, and equipment of every character, but no one, not even the guard, who so stoically performed his duty, thought to cover up these materials from the rains . . . They were soggy wet, . . . but . . . C 'est La Guerre!

To aching feet was added gripping pains in our middles. Mud bespattered boys, pulling feet slowly from a sticky, slime-covered road with a resounding 'sock' at every step, now began to move to the side of the flowing current of humanity and sit dejectedly on the roadside. The passing current sang at them:

"Arise you prisoners of starvation;
Arise you wretches of this earth,
For justice thunders 'damnation',
Gettin' slimmer 'round the girth."

The road passed along the lower levels of a hill. Here the barrage had been the fiercest. We looked northward for miles over a shell swept country and wondered how the machinations of man could make the destruction so complete.

Officers said we might eat. Tins of 'corn-willie', or 'monkey-meat', were ripped open. It was good . . . and we drank sparingly from our canteens, for

7 For a photograph of this hole, see THE FIRST WORLD WAR, Edited by Laurence Stallings, pp. 119-120. Simon & Schuster, 1933. I made the measurements of this hole the day we passed. CE.

we didn't know when we could get any more. A truck bearing commissary supplies passed slowly on the roadway. I ran my bayonet into one of the sacks and sugar flowed from the hole. My mess-cup was filled to overflowing, and I poured the sugar on the 'corn-willie'. It tasted fine. Holding it up, I shouted: "Sweet-William" . . . And "Sweet-William . . . Sweet-William" was echoed up the line of men . . . and the men kept running to the truck with their cups, and the sack of sugar was soon empty.

A faint sun-glow broke through in the west as if to shed some ray of hope and cheerfulness upon our progress into the Argonne Forest. We came to a place where a few trees were standing but their branches were snarled. We had to walk carefully on the ground for fear of falling into a shell—or fox-hole. Every few feet the earth would have a congealed red-splotch. This was the Argonne Forest! We 'fell out' to make ourselves as comfortable as possible. We had covered more than twenty miles that day.

The boom of cannon now rumbled out of the north like a deep growl. An occasional high-deflected shell whined its course above us, cutting a path in the sky, and breaking with a familiarity unpleasant and unassuring. Then some doughboy, to bolster up his nerve a little, would mumble:

"O, this battle of gay Parce,
Ifs making a lousy bum of me." . . .

The word was passed around from mouth to mouth that 'Elsie and the Kitchen' were lost again, "dam-mum!" . . . And there was no more "corn-willie" and "What the 'ell will we do?" An officer said he had seen some trucks park under the trees below us and as "you did pretty well foraging before we left the Saint Mihiel, would you like to take a crew and try it again?" . . . The drivers were already asleep out in the bushes when we arrived . . . We pulled back the tarpaulin and . . . Great Jumping Toad Frogs! . . . Joe Templeton's bald pate came poking out from under the blankets . . . "How in the name of did you get here?" . . . "Riding's better than walking . . . and it worked!"

Joe helped us unload two quarters of beef which had been frozen rigidly stiff before shipment. Soon thereafter little fires flickered under the densest trees whilst the boys of the company sat humped down roasting beef on the point of a stick. When night came,—when a flare would show the enemy our position,—all fires were put out again and we crawled into our pup-tents to keep off the cold air.

Garner, Big Tommy, and I tried to sleep together, but I hurt all over and couldn't sleep. I insisted on talking. They wished a "shell would hit you in the mouth . . . if it would only miss us!"

Daylight came. I tried to walk around but my temples throbbed. Black spots came before my eyes . . . and then after my head had gone in a whirl, I found Lt. Baker was holding me in his arms and asking "What's the matter?" He helped me to lie down. I felt better in a few minutes and wanted to go on again, but he said he had sent for the doctor . . . The doctor looked me over, took my temperature, and pinned a printed tag on my blouse on which he had made a check-mark opposite the letters "F. O. U. O."[8] He said an ambulance would be going back to the field-hospital after awhile. I must go with it and "be quiet" . . .

8 F.O.U.O.,—"fever of unknown origin."

Editor's Notes

Note on pages 113, 252 and 253 that Emmett mentions the incidental issue of pistols to many Soldiers of his company who were already carrying rifles just days before marching into combat. There was no training in how to load and fire them. There simply wasn't time. The weapons had just been delivered to France, the men were issued them as soon as possible, and they were to use their own initiative to figure out how to use them to fight. The mobilization of American industry and the Army logistics system were still trying to catch up with the burgeoning A. E. F. approaching combat.

The night before going "over the top" and attacking the German front lines in the St. Mihiel Offensive, newly-minted Sergeant Emmett and other leaders in his Infantry battalion were tasked to perform a leader reconnaissance of the route their troops would take again before daylight through abandoned trenches against the German advanced units. There are no maps and little information regarding where exactly the leading enemy troops were and the best way to lead their troops to get there. This was the reason why they were ordered to go forward, feeling their way along until they could hear or smell the enemy in the dark. Any noise that raised the suspicion of their approach would have caused an alarm and a sudden skirmish among the enemy sentries and the inexperienced Doughboy leaders. Their luck held that night. They paid for their bravery while leading their men the next morning.

U.S. Trenches at Monty Wall, Boullionville, France. Emmett and his comrades went "over the top" from trenches like these north of Fey-en-Haye, crossed No-Man's-Land, and cut their way through additional belts of German barbed wire obstacles under machinegun and shell fire. Though they started in the dark behind a strong friendly artillery barrage, the surviving Germans rose from the protection of their bunkers to defend their positions by the light of the shell fire and as the sun rose the next morning. Emmett was slightly wounded, gassed, and saw many of the men of his company fall around him during his first attack of the war. Stieghan Collection.

A propaganda postcard of a representative of the German Army that Emmett and the Doughboys faced in 1918. They had prepared for another war with the French and the Russians since victory in the Franco-Prussian War in 1870. These Soldiers were hardened, and their staffs and leaders were experienced. Though tired and outnumbered, these veterans were defending hard-won gains across Europe to protect their Homeland. Stieghan Collection.

Chapter 12

A Step Backwards

While I was on the ground waiting for the arrival of the ambulance Joe came and sat on a stump near me. We had lots of things to say to each other, but it seemed rather hard to talk that morning. He had a far-away look in his eyes, and I was wishing I didn't have to leave him, or that he could get a little pain that didn't hurt much and go with me.

While we were sitting there, just looking at each other, a pistol fired out under some bushes nearby and Joe went over to investigate. When he came back he nodded his head in the direction the gun had fired and said: "Remember that recruit you got when we were at the last barracks . . . the last stop . . . Well, he tried to load his pistol . . . and he's dead. Shot right under the chin . . . Bullet came out through the top of his head."

The vision then came back to me of my pistol lying on the table holding down his letter when I was mail-censor; and I was glad I had sent the letter on to his mother. I couldn't keep from thinking how unfair it was to send this boy up here without training. But perhaps, after all his letter would do some good back home!

Someone pulled at my arm, saying: "Shake a leg, now, Big Boy . . . We're going." I went along with the lieutenant and Joe toward the ambulance which was hidden under a nearby tree. They held on to my arms but I could walk all right. Joe asked me if I had heard anything of Bill, but he knew when he asked me I had not. The lieutenant said: "Goodbye; I hope you get along all right. Maybe I'll see you in Houston sometime" . . . I knew he was thinking I was starting in that direction.

I had to wait a few moments before getting into the ambulance while the driver arranged a place for a boy, who had become paralyzed in both legs and who had to ride lying flat on the floor of the car. After he was placed in the car I sat on one of the seats parallel with the length of the car and which were occupied by wounded and sick men. The paralyzed boy began to talk to me. I did not notice when the ambulance began to move and when I looked up I could not see Joe any more.

The paralyzed boy told us his name and said he had been all right last night, but had been under a terrific strain for a long time, and when he waked and tried to get up off the ground this morning he could not move either leg.

The following account appeared in the SAN ANTONIO EXPRESS the day after I arrived in San Antonio from overseas:

> "PARALYSIS DUE TO MENTAL STATE"
>
> Among the many and complicated cases handled at Fort Sam Houston is one in which mental condition had brought on paralysis. This man was brought into the hospital suffering apparently from total paralysis of both legs. The record of the case showed that he had been returned from the battlefield in that condition. There were, however, no wounds, no injuries of the spine, and what is more, the symptoms of organic paralysis were lacking. The limbs seemed normal, in good condition, except that the man could not use them. These facts, with others of the man's general symptoms, developed the medical opinion that the paralysis was functional, and that with a different attitude of mind on the part of the man the paralysis could be cured.
>
> But the man was perfectly sincere in his belief that his paralyzed condition was hopeless. He had been seized with the condition on the battlefield and now he 'knew' that 'he would be

> a hopeless invalid throughout life." The task before the physicians was to remove the man's belief in his paralysis, to make him know that he was well and could walk. This was accomplished through partial anaesthesia, at which times suggestions were given to the man and the man responded by standing up and finally walking. After his conscious mind was restored again, he was told of what he did while in an unconscious state. These treatments were continued until the man was finally convinced that he could walk if he tried. He did walk and soon recovered."

Our ambulance, with side curtains drawn, pulled slowly toward the west through the timber. When we found the road it sped southwestward. I sat in the rear to shift from time to time the position of the paralyzed boy, keeping an eye backward upon the expansive landscape now turned a dead-leaf brown. Puffs of smoke were rising here and there. A gray pallor lay close to the earth far to the north. Faint rumbles continued to grow fainter and fainter as we sped over a pitted roadway. When we had progressed into the depths of the valley our car turned so as to give my vision a clean sweep of the sides of a mountain. Dim, brown dots, hardly perceptible in their motion,—like small brown ants crawling forward in unbroken lines,—climbed higher and higher toward the crest while a thin grayish haze hung over their heads. Then while I looked, smoke puffs here and there fell into their ranks, and some of the little brown ants did not get up and go on again. A lone, plaintive whine swept off the mountain top and sang out into the valley. Its whirr made me think this long arm of destruction was straining once more to reach us. Then there was a terrible explosion above us; a metallic sound,—as if the steel gates of the heavens had been slammed for the last time,—thunder-clapped down upon us.

We came to a small tent capable of covering some twenty men. With no attempt to disguise it, it had been pitched openly in the valley. Its brown flaps swayed gently in the breeze. A stretcher was brought to our ambulance and the boy with paralysis disappeared. Other tags were read and the men unloaded. Finally an officer beckoned to me, but not before my attention had been attracted to the ceaseless pacing of an American sentinel just before the door of the tent. Up and down this doughboy trod with his bayonet gleaming on the end of his gun. He was keeping a wary eye on a German prisoner who also walked slowly up and down on the outside of the tent.

The flaps of the tent fell back and I entered. "Throw your gun and equipment there in the corner," and the captain looked at me, adding, "I don't think you'll need them any more."

In the corner were many guns, some with bayonets fixed, some with blood-stains showing a reddish-black on their highly polished blades. They were the discarded weapons of many men who had passed this way, perhaps never to return.

"Lie down: you'll feel better", said the genial officer. Then a shot rang out at the very tent-entrance. The captain pulled back the tent-flaps to look . . . and went hastily outside. When he returned, with a wry grin, he told me: "Yes, he shot him . . . the German prisoner! . . . Because he ran . . . But of course he ran . . . So would I . . . had he stuck me in the back with a bayonet . . . Yes, I found the bayonet-wound, fresh bleeding, in the German's back!"

Again I was loaded into a ambulance, and night came. There was a long, long ride. It seemed to have no ending. The two Frenchmen with their pointed mustaches, who went along, were drinking repeatedly from their canteens and talking unceasingly. One had been shot through the arm. His arm had been broken. The blood oozed through the bandage, but he shrugged his shoulders when I asked him about it, and answered as if he were surprised I thought the wound might hurt him: "Combien? C 'est La Guerre!"

When we stopped late at night before the dim outlines of a stone-building, I jumped to the flagstones and would have fallen but for the arms of an attendant . . . "Not so fast, buddy. We'll carry you." . . . "Not on your life . . . I haven't lost a leg," but when I tried to walk my knees failed, and laughingly, they took me by the arms and I was soon in a long hall. Steel-helmets,—rows of them,—rested on a table paralleling the wall. Onto the tops of them stuck tallow-candles which threw a pale-yellow light.

A doctor was bending over a bleeding man. Another was swearing because the attendants would not keep the bandage drawn tight across their noses. Another looked at my card . . . "Through that door, then to the left", and I sat down on a cot under a tent which had been spread to connect up with the big stone building. My tongue was feverish. My mouth was dry. I tingled all over. Acrid fumes stung my nostrils. I tried to relax . . . and distorted figures ebbed and flowed before my eyes . . . Then all was quiet for a long time.

Someone tugged at my arm . . . It was daylight again . . . And I looked up into the face of old John O'Bar,—L Company O'Bar. "What the !How'd you get here?" . . . "Hurt in the back coming into the Argonne . . . Been here two days . . . Heard your name called . . . Let's eat!"

We walked out into a slow falling mist, but when the aroma of the kitchen came my stomach revolted. We sought shelter across the street where we saw a Red Cross canteen sign showing on the side of a building. We entered . . . "My gosh, John. Am I seeing things?" I pointed to an American woman, the first one I had seen since leaving England . . . She smiled her inquiry: "Am I such a sight as that?" . . . Had I been sanely frank I might have been affirmative in my agreement . . . but she was sociable and offered chocolates. I declined, saying "Big Bertha has been throwin' 'em up pretty high this morning, so I'd better not load up." Then she handed me a lemon with the explanation that it was "not symbolic", adding perhaps I had better try my bed the remainder of the day.

When I arrived at my bunk an inquiring medico-captain was wondering if I had gone A. W. O. L. He thumped my chest, took my temperature, probed around in my throat, peered into my nostrils and suggested "You'd better be evacuated". I told him this had already been successfully accomplished; that I had only this morning spit up every breakfast I had eaten in three months. He put his hand over his face trying to press back a smile and said: "Well, we'll evacuate you to another hospital, anyway!"

A dinky little engine backed many cars into a side-track near the hospital and stretcher-bearers worked rapidly for a long time carrying men covered over with white sheets aboard the train. I walked out and climbed into a designated coach. It had three compartments. On the floor sprawled many men. The seats were occupied, also, but a man who could stand up made room for me. We shared one seat together. The train pulled out and ran a short time when night came again, and we were left with our thoughts and the groans of men.

All night long the men on the floor moaned, groaned, swore, and prayed. The boy who could walk went with me and we gave the sufferers water, rolled blankets into pillows and tried to make them as comfortable as possible. The train stopped in a large city. A nurse came to the train to ask how we were getting along. We told her we had two pretty sick boys but one had "become quiet now". She came into our compartment, pushed back the boy's eye-lids, felt his pulse, covered up his face with a blanket, and told two soldiers standing outside to bring the stretcher. They took him away with them. Then she said she'd take the other fellow, too.

There was room now for two more men to lie down. Two of the badly injured men lay down on the floor, but this did not make room enough for me to lie down, so I climbed up in the luggage-basket above the seats. The basket had an iron bar across its middle to support the luggage. The bar cut into my back and I had to turn over and let the bar rest across my stomach. This position was a little more comfortable, but I couldn't sleep. We rode all night. When day came they told us we were entering Paris. Our train stopped a long time, then backed up, and headed southward. We rode all day again and far into the night before we stopped. This time we were told to unload, and we stood near some buildings, which in the darkness looked like the cantonment buildings back at Camp Travis. I walked about a hundred yards to a building entrance. A captain met me at the door and said I'd have to take a bath. It had been a long time since I had taken a bath, but the water was warm and it made me feel comfortable all over.

A nurse showed me a bed . . . It had white sheets on it . . . but I was now hot and burning all over, and the cold clean sheets felt so comfortable.

I wanted to sleep, but a man dressed in white came to my bed and hammered on my chest . . . Big eyes were staring at me . . . Then the same man put his head to my chest, listened, and said: "rales".

It seemed a long time afterwards before the weight on my chest was lighter. Then I opened my eyes and a dour-faced nurse was pointing a finger at me: "He won't eat, captain." The captain laughed and asked: "Feel better, old fellow?" He sat on my bed and talked a few moments while he felt my pulse; then he placed a stethoscope to my chest, and started away: "Coming around fine. Will drop in for a chat with you this evening."

The boy in the next bed said I was in a hospital near Nevers, that it was now November second, and that "whatever I didn't want to eat to give it to him" and . . . if "you don't recall . . . my name is Ferguson." I told him he could have everything they offered me to eat from now on out "to the end of the emergency."

The dour-faced nurse caught me giving Ferguson my supper and I was reported to the captain. The captain came and sat on the side of my bed again and asked if I thought I could eat some soup. When the dour-faced nurse said they didn't have any soup today, he told her to fix it anyway. He wanted to know where I lived,—but all the time he was listening through his stethoscope while I talked. When I told him I was from Texas, he quit listening at my chest and said he had a good friend there named David Picton. I told him Picton and I had been school-mates. Then he told me not to pay any more attention to the woman with the dour-face and he'd see I got along all right.

A night-nurse came to our ward about this time, who smiled and whistled all the time. At night when everything was quiet she'd made candy over an alcohol-burner for those of us who could eat it. Often times she stole lemons from some place and we had hot lemonade. After she came the men's spirits were higher, and they would sleep in the day-time so as not to have to look at the nurse with the wrinkled face and so they could stay awake at night to talk to the cheerful nurse.

Whenever things seem to be going the best there is always something or somebody to 'mess up the detail'. The pinched-face nurse told the major our ward needed a careful supervision. A corps of 'brass-hats' came through looking over everything very carefully and questioning everyone. We had a young Italian boy in the ward who had put his foot down on an exploding bomb. He was now recuperating enough to walk around on the stump of a heel the doctors had kindly left him. When the 'brass-hats' came around to him he asked the major: "How far is it to Italy from here?"

Such a significant inquiry naturally indicated to the minds of the visiting officers that the boy was soon to be, with favorable conditions, homeward bound. The result was: A guard was placed over the ward under specific instructions to keep constant observations on the limited movements of the little Italian.

A furore of discontent broke out, pervading all the patients, because of the repetition of the statement by disgruntled patients that "you can't be a wounded patient in this man's d army without having a bayonet pointed at you while you lie helpless in bed."

All this time the Italian with the bad foot would call out tauntingly to the orderlies: "Duck", . . . "Duck; orderlies, breeng de duck". Then when the 'duck' was brought to his bed he would declaim his inability to use it, only to repeat the call in a few minutes.

I talked to the genial Irish lad from New York across the aisle. He was telling a good story when his lungs began to hurt him. He'd finish tomorrow, he said,—when he felt better. When morning came I looked across to his bed. I no longer could see his jet-black hair above the white pillows. The nurse saw I was looking for my friend and she put a finger up to her mouth and nodded her head toward the front of the ward. There I saw a white curtain drawn across the corner of the room. I could see two beds had been moved behind it. Nurses stood around and peered behind the curtain all day. Officers came and went, stepping very softly all during the night, but when morning came the curtain had been removed and the beds were vacant.

They transferred our genial nurse to another ward, and we could see the woman with the wry face was pleased . . . After that all of us had a sudden resolution to get well as quick as we could and go back to the front . . . or anywhere . . . This was no place for sick men.

My medico-friend came to my bed again . . . "It was unfortunate about your night nurse . . . Best we ever had . . . but C est La guerre . . . I have concluded you need an operation, but for the present your nose and throat are too badly burned by gas. Ether would give you pneumonia again . . . I'll mark your record so the 'bone-chiselers' will keep off you." . . . He wrote on my record: "Pneumonia induced by gas inhalations. Operation needed: Not recommended, dangerous." . . . If you go back to Texas, tell Picton you saw me."

Chapter 13

The Battle of Monotony

My 'deloused' clothes were now returned to me, and I stepped out into the mud covered hospital area. Before the door a large group of German prisoners were digging in the ditches and leisurely throwing out the muck to drain the sodden streets. American boys carrying reeking, brimful cans were careful to splash some of it onto the prisoners as they passed. An expansive hospital area stretched down the hill-side. I had been quartered in the first ward adjoining the railroad tracks. Across the tracks floated the flag of the United States. It tugged energetically in a cold breeze against its halyard.

I sat in the drizzly rain before the headquarters waiting for a sergeant who finally came and formed us into a line. We marched slowly southward. We wanted to go faster to get out of the rain and the cold wind, but our legs were weak. We came to a large frame building. By that time it was night, and I sank down on a wire cot. I couldn't go to sleep, so I found the supply sergeant and he issued me two blankets. After that I kept warmer. The stamp of many feet treading on the wood-floor beat upon my ears far into the night.

I awoke late the next morning. There was no bugle call, and the men in the cot-filled house rolled fitfully on their beds. I decided to walk around. Reporting to the orderly room, I inquired: "Will there be any orders today?" . . . "Wait around and see!" . . . "O, yeah?" . . . and I walked off toward the railroad tracks . . . A Frenchman with hammer in hand, who seemed to be in a great hurry, stopped before the wall of a mammoth warehouse to tack up a bulletin. I walked over to see what it was. I read:

"November 11, 1918.

All hostilities ceased today at
11:00 a. m. on the Western Front."

I walked away . . . dazed . . . A soldier approached me . . . "What is it?" . . . "It's over . . . Read it" . . . He stared at the bulletin on the warehouse wall, then threw his 'tin-lid' high in the air and yelled . . . Men crowded around the placard . . . They ran away shouting . . . Frenchmen screaming: "Fini La guerre . . . Fini La guerre!"

I sat upon a bench to watch excited men fall into marching lines. Some hobbled on crutches . . . Some carried their crutches under arms . . . Some threw away these appendages and stamped off without a limp. A roll of drums boomed out from the village below the evacuation-area. Line after line of men merged. The roadways became one continuous mass of moving men, the khaki interspersed with the blue tinged with red, brightened here and there with a nurse's uniform. On and on they flowed . . . "Thump . . . te-thump . . . thump . . . thump . . . thump . . . thump", and the cadence was drowned by the singing throng:

"Over there, Over there;
We're going over;
And we won't come back,
'Till it's over, Over there!"

My weak legs failed to respond to my desire to join the throbbing, clamoring mob. I turned back to the orderly room where I was greeted with a salvo of curses all of which taken together created the impression upon my mind that I, by my absence, had 'messed up' some examining-medico's well laid plans for the day . . . "Where the hell have you been?" . . . "And who in the wants to know? . . . You don't seem to know much; not even that the war is over!"

The sergeant's eyes popped out, deflating his simulated importance to normal, reminding me of an icicle subjected to a sudden burst of heat . . .

Then he asked: "Is that what the noise is about?" ... "If you were not so intelligent I would say 'Yes.'" ... With "Report at 1:30 to the infirmary", he grabbed a 'Mellon-Special' (indulging a futile hope it would keep off some of the rain) and ran toward the marching gang.

At 1:30 a familiar face appeared with me at the appointed place. My name was called. The man behind me asked: "Are you from Victoria?" ... "I was, one time." ... "I am Evans from El Campo." ... and the doctor admitted us into his office together ... "Operation not recommended, eh?" ... "Well, we'll evacuate you tomorrow to the Nevers Convalescent Area" ... "And you, Evans?" ... "Shot under the eye with rifle bullet", he read from his record.

The examination of Evans confirmed the diagnosis that he had been struck by a bullet which entered his left eye socket. The bullet had lifted and pushed the eye-ball away. It had then taken a course through his body lodging against the spinal column on a line with the fifth rib. His medical record showed an operation had been performed and the bullet extracted. The sight of the eye had only been slightly effected. No paralysis had resulted.

Evans explained the peculiarity of the injury in this manner. He was lying behind a spool of barb-wire on the front exchanging shots with a German sniper. In order to take a better shot at the sniper he exposed enough of the side of his face to see down his rifle barrel. Then something stung him just under the eye ... The next he knew was when he revived in the hospital many weeks later ... The course of the bullet had been with the length of his body because of his prostrate position.

The eternal din of marching, yelling, screaming people ebbed and flowed in still greater and greater intensity as the night came on. It seemed all humanity had turned vocally hilarious. While the hubdub floated indoors to us, a wounded negro soldier, evidently from the South, walked slyly into the spacious sleeping quarter and picked out a cot near me ... "Not next to me, Black Boy" ... "If 'tis all right with you, Boss, I'd sorter like to get close to dis door ... Seems like when dem boys come in drunk tonight I gonner have to leave here sorter sudden like, and I'd shore like to have dis door convenient . . ." He and I moved my bunk far back into the hall and the black boy availed himself of the convenience of the exit.

Faint streaks of light broke through the east. Unsteady feet sought deserted bunks. Drabbled 'wraps' were unrolled and cast upon the floor. The din of a night, the like of which there has never been another, had come to an end. Sore throats ceased to utter sounds. Cognac-glutted bodies lay down to rest. Buglers failed to call forth eternal curses. November twelfth—the morning after—had come!

A barrack bag was issued me. I sought out my new-made friend, Ferguson, and carpet-baggers like, we waddled on feeble legs to the Nevers Convalescence Area. The prospects were uninteresting. A 'British tent', capable of housing forty men, but now crowded with twice the number, became our abode.

Indolence was the order of the day. The inevitable reaction,—from hilarity of spirit to remorse, rapture to contrition,—settled down upon the camp. The Spirit of America, far away from home, was at a low ebb. It was over. We wanted to go home.

Line after line, row after row, the canvas-city spread over the water-soaked prairie. A restraining barb-wire fence encircled the camp. Sentinels stoically walked their posts. The rains beat down upon them. Winter had set upon us. Egress from the camp was permitted only on passes which were not obtainable by casuals . . . And we sank down upon wire-covered cots, hoping for, but not expecting, a rest. The drive of water-laden winds converted tent-floors into mushy quagmires. Cot-legs sank down into the mud, and our beds went flat on the ground.

A long period of inertia brought a decision to Ferguson and me. We would 'go over the hill' for a visit in a nearby village. We effected an exit, surreptitiously of course, and were soon within the portals of an 'estaminet', behind the protecting 'Off Links' sign where a French girl served us a delectable dinner. At her skirts tottered a babbling baby. I spoke to her: "Madam?" . . . "No! . . . No! . . . Yank . . . Me madamoiselle!" . . . "But whose picaninny?", pointing to the child . . . "Madamoiselle's, Yank . . . Ees not one picaninny blenty nuf for madamoiselle?" . . . We thought so.

While homeward bound we joined an excited group of Frenchmen and Yanks who looked with a voiceless woe toward the rafters of an abandoned house from which swayed the dead form of a suicided American soldier . . . "Well, Ferg., he was not so crazy after all . . . He left this mud-hole didn't he?"

Nevers lay to the northeast of the camp. Our previous success in 'sentinel-jumping' inspired a new trip. We, therefore, determined to go to Nevers. This time the tender of a package of the 'makin's' was sufficient for the guard to show us a convenient hole in the fence. Through it we made our way into the village. Hardly had we gone upon its streets before we were accosted by a Frenchman with tonsorially perfect pointed moustaches, who inquired of our social acquisitiveness . . . "Who?" . . . "My wife and my daughter! See!" . . . And he pointed to an old woman and a chic madamoiselle who appeared from out the darkness . . . Other soldiers joined the group . . . and it was not long before 'her chicness', armed about with khaki, disappeared into the night whilst 'Immaculate Whiskers' gleefully counted the francs left him by the departing soldier.

As they faded into the darkness the lines from Othello came back to me:

"Most potent, grave, and reverend seigniors,
My very noble and approved good masters,
That I have taken away this man's daughter,
Is most true." . . .

"But you cannot", said Ferg.," complete the speech with truthfulness:

"True, I have married her;
The very head and front of my offending,
Hath this extent, no more!"

"Well anyway, we'll see this through" . . . and the two of us followed behind the Frenchman now smickering in his venality. He entered an 'estaminet' and purchased a bottle of wine. Mounting a table he began to sing the "Marseillaise" with voluble gusto . . . All stood at 'attention' . . . then . . . "crash"! A shattering bottle broke over his head . . . He tumbled from the table . . . The M. P's. whistle shrilled in the street . . . Ferg. and I sauntered along nonchalantly (lest we be suspicioned of attempting to slay the 'Frog'), then ducking into an alley we ran for the 'hole in the fence' . . . and slept in camp that night . . . If the Frenchman's head was sore we did not care.

With the relaxation of spirit, after the Armistice, a throbbing religious fervor spread throughout the American army, now impatiently waiting to go home. 'Church Call', unrecognizable by American ears before the Armistice, now drew slow moving crowds toward places of worship. Y. M. C. A. huts (although there was a rancor in the hearts of nearly every red-blooded American against the effete Sisypheans in charge,) were filled to overflowing. Men discussed home, friends, wives, sisters, and mothers, accrediting their Maker with the good fortune of their survival.

"On ward Christian so old jers
Mar ching as to war,
With the cross of Je sus,
Go ing on be fore."

The chaplains taking advantage of the pervading fervor, a spirit which,—try as hard as they might,—they had been unable to instill, admonished home-sick boys to remember the blessings of their home and write mothers and wives. Steadied by recuperating strength, and now thinking of the days to come, I wrote retrospectively:

"Near Nevers, France,

December 9, 1918.

"Hon. ------ ,

———————, Texas.

My Dear —:—

I don't know whether I ever put my good intentions into a deed and wrote you or not. I hope I did. So many things have occurred over here that I fail to recall even the most important ones,—hence my mind fails to record whether or not I wrote you of this country. Imagine yourself relegated to the Middle Ages, surrounded by the hords of old, by soldiers, by guns, by all the man-killing machinations conceived by the evil mind of man, and put all these in mud, shoe-top deep, with a lowering sky, and you have a faint conception of France.

Dad has likely told you that I have seen my part of the hell of the front. It was that and then some. I was in the big drive at Saint Mihiel and drove from Fey-en-Haye toward Pagny on the Moselle to within twelve miles of Metz. (Don't ask me why I did not take Metz!) I was also in the Argonne, later, but did not go over the top there with the boys. I took sick with bronchitis and was hauled away. I am now expecting to be sent into Germany where my Division, the 90th, is located.

" . . . there is no way to describe what has occurred over here. You would think from the number of O. D. suits you see that all America has moved across. I am ready to make the return trip, though, and we all hope the time will be short here. We feel we are entitled to a return to America as soon as conditions will permit. We have tried to do our part, and now want to take up our several vocations . . .

"I have seen all the misery of war that I want to see. Men have been shot down by my very side, and I was listed as one of the fortunate ones. A man is not wholly human when he is in a fight. I experienced no sense of fear when in the fight, but it is always hard to get your nerve up when lying ready for the attack.

"I shall never forget the morning of. September 13, when, soaking wet, half frozen, I lay in a mudhole, and listened to the shells hissing over me. When we got going it was easy.

> "I fell into a shell-hole on top of one of our sergeants and he was laughing, and said: "Emmett, old boy, this is what I call fun." We felt just that way. There was no danger felt by us, and even the death of our friends around us failed to impress us. When once we stopped, though, of course, there was a relaxation" . . . "

A complication arising out of the condition of my throat, yet far from well, threatened for a time to impair the sight of an eye. Thanks to the solicitous care of a captain from Chicago,—who had a brother who was a major and who said 'he practiced medicine while his brother practiced polities',—I was put well onto the road to recovery, but he invalided me and marked me for 'home'. I was to be shipped via Saint Aignan.

The ride to Saint Aignan was via the famous '40 and 8', which necessitated our being again run through the 'delouser' before we could be assigned to tents. The 'delousing' process took place in the middle of the night. Day approached before we received our assignment which was in a tent in the valley across from Saint Aignan. I laid down my barrack-bag and stepped out into the company-street before my tent, not knowing of the order of the camp prohibiting one being out of the tent after "taps". I was immediately arrested and thrown in the 'brig' where I remained for two days. Strange to say, I had little resentment over the incident. When I refused to request a 'trial' they liberated me.

I had been told by the other prisoners that the camp was being dominated by a captain named Smith. (I often wondered if he could possibly have been the far-famed Hard Boiled Smith who gained such notoriety in a congressional investigation just after the war!) This officer, so the prisoners said, had been iniquitous in his treatment of the troops flowing through the 'rest' area.

The food was bad at Saint Aignan. The beds were a minimum, consisting only of blankets laid upon the ground. A tributary of the Loire river, which flowed bank-full through the valley throughout the winter, was spewing its scourge of muddy slop over the tented area. Winter had set upon us in its fierceness. A stove in a tent was a rarity, indeed. And when the river oozed its muck over the bank to spread out inches deep in the lower reaches, (some of which flowed through my tent,) we bartered with the night-guards to let us pass and stole saw-dust from the 'Frogs' to levy back the creeping scum.

Day after day a line of soldiers marching eight abreast struggled to get to the kitchen which stood under a mildewed tent across a road which ran through the middle of the camp. Many men finding themselves unable to cope with the weather stayed in their tents without food or shared a ration with an obliging buddy.

The tents were built to serve eight men. When I arrived ours already had eleven men sleeping under it. One day a flap was raised and a sergeant inquired: "How many men in here?" . . . Figuring we were to have another addition to our 'happy' family, I answered: "Fourteen," and he passed along. I listened, and at the next tent I heard a familiar high-pitched voice. I went out to investigate . . . and from out the adjacent tent crawled Bill Goodson . . . And of course we had room for one more!

Feeding five thousand sick and injured men from one kitchen was not the easiest of tasks. Five thousand body-shattered men with nothing to do but grumble and "want to go home" created a problem, also, most difficult of solution. Fights were frequent, especially in the 'chow-line'.

Attempting to interpose himself into the situation, the captain of the arrogant imperiousness located himself at the roadway on a bridge over a slough where all men must pass while 'marching unto chow!' He stopped and started the flow of men as the exigencies of the situation at the kitchen required. His demeanor in performing this service was such that no man ever recalled him with pleasure. The violence of his epithets was equaled only by the daring with which he blurted his words. One evening the line moved toward the kitchen slower than usual, it seemed to us, in a downpour of rain.

Hungry, wet, shivering sick and wounded men were irked beyond control by his profanity. Under the bridge, some three feet below his feet, flowed the filth and scum of an oozy camp. The file of men flowed on . . . Bill Goodson approached . . . A mess-kit rattled off the head of the captain and 'his dignity' fell full length backward to disappear below the slime of the filthy gutter. Men scurried away in all directions. I ran forward and Bill caught up with me. He placed his finger over his mouth to indicate 'silence' . . . And, investigate as they might, army officers learned nothing from the mistreated men at that camp about the 'down fall of the captain'.

Christmas Day came. About one mile west of us was a larger kitchen capable of feeding the entire casual camp. Orders were issued that all men should remain in their tents until we should receive a printed 'greeting' from the Y. M. C. A., which 'greeting' if presented to the area hut while en route to dinner, would entitle each soldier to a Christmas box. E. F. Davis, sick in our tent and unable to walk to dinner, asked that I receive his box for him. I presented myself at the Y. M. C. A. hut just as the attendant, resplendent in his simulated soldier's attire, came to the door to explain that he was closing to take Christmas dinner with friends. I asked him to deliver the sick man's package . . . "Sick, or well, he will have to come here after it" . . . and the Y. M. C. A. again ingratiated itself to the 'boys over there'.

I found the 'chow-line' at the new kitchen a long one. Back of me a soldier was creating quite a commotion by his loud talk. Later some of the boys told me this man's chief occupation in life seemed to have been gambling and drinking. A 'hard-boiled' sergeant stood at the head of the file to direct the serving. He, too, had a reputation for uncouthness which had spread throughout the Convalescent Area. After I had passed beyond the 'ladders' I heard a commotion. Men were running in all directions. I knew there had been trouble, but I did not go back to see, but the men said the mess-sergeant had fallen with a mess-knife, sharpened keen on both edges, planted deep into his heart . . . Many days afterwards, rumor had it again that the soldier with the vitriolic tongue had been arrested for going A. W. O. L., but the Frenchman, in the upper story of whose house he was found, explained that the soldier had been drunk there for many days.

Down the valley and across the bridge-head from the city of Saint Aignan the Y. M. C. A. had established a recreation center. The art of fistic encounter was the chief source of amusement for men who had been steeped in turmoil for so many days. Davis, Obar, and I decided to attend a fistic-session. Bill was to join us later.

Upon our arrival the hall was almost empty. My strength was rapidly returning now after my illness, and I engaged another soldier in boxing until required by the boxing-master, a product of the 'side-walks of New York', to quit. I then sat down in the front row of seats. My attention was attracted to a gangling, yellow negro, with 'over-seas' cap stuck contemptuously on the side of his burly head, who was making his way along the wall toward the stage. I remarked to Obar that "should the 'bur-head' try to get into the game some one should decapitate him with a properly seasoned club". Being from the South, Obar agreed.

The house had now filled to capacity. There was space left for the late-comers only in the windows. A lull came in the bouts and I jumped back on the stage for another round without so much as looking to see who might be my antagonist. With gloves tied on, I turned to be shocked: I was confronting the yellow negro.

At that time I was unsuccessfully cultivating the growth of a moustache. The boxing-master, with one look at my face sensed my feelings. He said: "I see that the gentleman 'with the misplaced eye-brow' is from the South . . . He will not fight the negro." . . . "No, of course, I will not!" . . . "Who then will you fight?" . . . "Any man in the audience (And immediately I feared I was covering too much territory!) who chooses only men of his own color for playmates!" . . . and the crowd both yelled and hissed . . . "Name one then!" . . . I faced the seething crowd of men. Far back in the hall, sitting precariously in a window-ledge was Bill Goodson. It was plain he was enjoying my embarrassment . . . Pointing, I called: "Bill Goodson" . . . He climbed down, and with his long arms swinging, trotted toward the stage, and I knew I was in for it!

He hammered, slashed, and beat me, his long arms reaching me at will. But I determined to win that fight, and summoning all my endurance withstood his smashing jabs. Hoping my chance would come, I saw him gather his strength just as my endurance was fading. I knew he intended to finish me with one more blow. Then I smashed at him with every ounce of strength I had left . . . The blow landed squarely on his jaw . . . His feet went out from under him and he crumpled into a heap . . . The 'gentleman from New York' poured water on him, while I slipped off the gloves and departed. Late that night Bill crawled into bed with me, complaining: "Well, why didn't you kill me?" . . . "It had to be done, Bill. The negro-question was too prominent". He agreed, but contended it was terribly hard on chins.

Across the river was a gem of medieval grandeur, the castle of Saint Aignan. The rugged uplands broke abruptly before the river. On the very pinnacle of the heights,—a position elevated, as if designedly by the hand of God to command a sweeping view of the measureless valley below and the sylvan hills beyond,—stood this intriguing chateau. Skirting its lower reaches a forbidding wall obscured its portals with dignified beauty. A fortress once, but now it was symmetric elegance which beckoned me to explore. But an acrid Frenchman said: "No."

And I stood there upon the heights of one of France's many castellated hills, the cold, age-browning stone-walls rose majestically behind me. Below, in the valley, flapped the tents of the great American Casual Camp, row after row of pale brown measuring across the valley. Mountains of a second rate, to the south and to the west, held their heads above the obscuring smoke rising from the camp. And as I looked the old Frenchman spoke: "The Spirit of America now! . . . Once it was the Spirit of France! . . . Joan of Arc once rested there . . . Into that mountain (and he pointed far off to the left at a small scratch on the mountain side) fervently obedient, under the spell of her invincibility, her hords burrowed winter-quarters" . . . I would see that, too!

The turgid waters of the slimy river trickled here and there through apertures emptying into the lowlands. Slowly but surely it encompassed our camp, and when we had reached our tent that evening the ditches along the roadside had filled to over-flowing.

Morning came and there was a field of tents set down upon a boundless lake. Deep pits had been dug beneath the rudely constructed *cabinet d'aisance*, and the flood had caused them to cave in. Two boys, racing in the early morning to perform their lavational requirements before mess-call should sound, unbalanced and plunged headlong to disappear beneath the noxious abyss . . . and an order came that the camp would move to higher ground!

We walked in the water, and soles of water-soaked shoes, weakening under their flaccidity, separated from their uppers. Hundreds of men stepped unprotected in the slime. But the army order was "that the camp had been properly equipped and no equipment replacements are permitted" . . . Then when we moved from the sodden camp our course lay alongside a salvage-dump. There, stacked high and unprotected from the weather, we saw multiplied thousands of unsorted shoes!

Our retreat from the valley to the gently sloping hillside to the south was greeted by the swerving of the wind and its blasts beat down upon us from the north. The mud-field crusted over. Snow began to fall. Rumor came, lip to lip, that we were going home. A buzz of excitement arose above the blustering gusts of wind. Men ran to the bulletin-boards to see if "embarkation-orders' had been posted. Then in disgust, the songs of penance subsided to be supplanted by roisterous tones, marking the passing of the religious fervor.

Hopping dice amused the men as they sat around spread blankets calling plaintively to "Little Joe" . . . "Come seven" . . . "Baby needs a pair of new shoes". Kneeling men peered over the players' shoulders rimming the 'galloper's course'. "African golf" and "Pershing's Delight" had invaded the chaplain's sanctum!

"Little Joe" dealt unkindly with me and I found my 'reserve' reduced to one franc. I decided to wander from the camp, regardless of restrictions, and see the Cave of Joan of Arc. A bracing 'shot of cognac', surreptitiously acquired and planted under the ribs of the guard who trod the snow-covered post, was sufficient to cause him to tarry at the farthermost end of his beat . . . and we were soon across the valley and climbing up toward the mouth of the cave . . . A two wheel cart piled high with hay rolled out of the cavern's mouth. We entered and were swallowed up in darkness.

The cave, according to the legend conveyed to me by the Frenchman, had been dug into the mountain by the army of Joan-of-Arc who used it as winter-quarters. The transmutations of passing ages, however, had converted it to modern usages, and we found it the convenient storage place for the Frenchman's ever-abundant liquor supply, which he bedded down with his winter forage.

We found two entrances to the cave in the semi-circular mountain. From mouth to mouth, measured across the valley in a straight line, the distance between the entrances was about one mile. The channels which ran into the mountain from an approximate level with the valley were wide and capable of permitting the passing of loaded wagons.

Light within the cavern was limited, for it came chiefly through holes drilled from the top of the mountain intersecting the myriad of channels below. After our eyes had become accustomed to the darkness, we walked apprehensively along the passageway to enter chamber after chamber. High arched ceilings cut into stone stirred our imaginations. Industry born from necessity had left behind an entrancing handiwork.

In the darkness of connecting channels we became confused. We wandered aimlessly in the bowels of the earth unable to determine our direction. We were lost. A surge of hysteria encroached upon us. A desire to flee into the unknown obsessed us, but a steadying voice demurred: "What's the hurry? . . . It's warm in here!"

Voices echoed faintly down the hall, and we followed to them, finding ourselves in an immense chamber. A shaft of light passed down to us from above. And men, walking slowly protecting lighted candles with cupped hands, descended eerily upon us . . . "Take a candle, buddies . . . Watch the arrows on the walls . . . They point out." . . . and another party of Americans glided deeper into the bosom of the earth. We looked after them as they passed down the phantom halls, their flickering lights casting impish shadow-sprites about them, and I thought of the "Alciphron":

"Wonders on wonders; sights that lie
Where never sun gave flow'ret birth;
Bright marvels, hid from the upper sky,

And myst'ries that are born and die
Deep in the very heart of earth!"

As we walked along the cheering light which came down the light-wells became hazy. We knew the day was waning, and sped up our faltering steps toward the exit. A white mantle broke before us,—the earth covered with snow,—and we were outside again. In the far distance the yellowish brown tent-sides, tops covered with snow, gave us a sense of tranquility and peace . . . We returned slowly, satisfied with the day.

Approaching the camp, "tap" . . . "tap" . . . "tap", the eternal hammering of francs, filtered out to our ears. A peculiar hobby had invaded the ranks of the impatient American Expeditionary Force. The franc, having little value as a coin in the eyes of the American soldier, was a sufficiently soft metal to be turned by gentle tapping on its edge. Constant thwacking on the upturned edge of the coin flared out its rim so that the words, "Liberte, Egalite, Fraternite," appeared beneath the rim. When the center of the coin was then cut away, a French ring was the result.

Multiplied thousands of francs were hammered into souvenir-rings each day. So serious did the coin mutilation become that a general order was issued prohibiting the mutilation of the coins of France. It began as a hobby so that sweethearts and wives back home might have a souvenir, but it actually threatened the spoilation of the monetary system of the republic before the boys moved away.

A new recruit to our casual company came about this time. A finer lad,—a gamer sport,—never donned the uniform in the cause of liberty. Small of stature, keen of mind, but unkempt in appearance, he rivaled the best in the art of inveigling 'Little Joe' to do his bidding. But deft as he was in the acquisitive art, his riches would depart as soon as he relaxed 'to celebrate his prowess'.

One cold night when the snow nestled against our staked-down tents, I heard voices outside. A dull thud followed. Shortly thereafter the tent-flaps raised and he, who in the darkness resembled the soldier whom rumor had it was nearby when the mess-sergeant was killed, passed quietly through our tent to disappear by the farther side. I investigated and found 'Fortune's favored one' lying "unhorsed, lame and impotent" against the tent buried in the snow with a crimson mark across his face. He could tell us nothing other than he had 'won' that night. When morning came he inquired for his blouse. We found it outside buried in the snow with the pockets turned wrong-side out . . . He took it with a smile, adding: "I'm not 'busted' yet". With a knife he cut away the stitches and from the inside of the lining extracted a handful of paper money of large denominations . . . "They didn't get that" . . .

Finally, the long expected call came: "Outside with barrack-bags" . . . and in a falling snow we mounted 'side-door pullmans' to roll away toward the southwest. The biting cold whistled through the unheated cars, and freezing men cursed and whimpered.

Darkness came . . . and . . . an interminable ride, . . . then morning. Our blood warmed under the inspiring scene which skirted our course as we entered the valley of the Dordogne, that river of magic, hedged in with jagged peaks and festooned with chateaux, the stolid magnificence and dignity of which rendered men inarticulate. I stood in the doorway of the car and looked above me to the heights. There stood fortresses of old, things of scraggly beauty now,—implanted high above the splashing river with sloping terraces, garlanded with the dark green of winding winter-vines . . . And we were over the mountains . . . entering Bordeaux.

Our commander, a newly assigned wounded aviator, had suffered haplessly during the night in his efforts to dispel the cold. His feeble legs failed to support him when we drew up at Bordeaux into formation. A less unstable officer assumed command, and we rode away to a nearby camp to 'delouse' again.

Here we were ordered to pour out the sugar from our condiment cans. "You are going home and won't need it again." A steady file of men passed all day long beside a single barrel, each man depositing the sugar from his can. It began to rain; no one thought to cover up the sugar. Then the barrel filled, and sugar ran out onto the ground. The men marched through it with their muddy feet, each one pouring more sugar onto the flowing pile . . . because that 'was orders'. But such were the ways of the army!

Then they ran us through the 'delouser' and my clothing was lost in the shuffle . . . And I wanted to go home, but it looked like it would take an act of Congress to provide me with proper apparel . . . But that, too, was the way in the army!

A few very pleasant days at the camp near Bordeaux passed swiftly while we trained our ears to catch the hoarse, throaty blasts from the ships plying in the nearby Gironde River. Shrill, piping, tug-boat whistles told us that ocean-liners were being shifted from their berths to the stream . . . And we urged stagnant time to relieve us from this Battle of Monotony,—take us along!

Once more we slung our barrack-bags across shoulders and marched. We passed along an endless road cluttered with trucks, and the drivers shouted to us: "Give 'way to the right". Then we looked down upon bristling masts standing before a long shed . . . Men were marching, disappearing within, and we sang:

"There's a long, long trail a winding,
Into the land of my dreams,
Where the night-in-gales are singing,
And the wide moon beams,
There's a long, long night of waiting
Until my dreams all come true,
Till the day when I'll be going
Down that long, long trai . . il . . to yo . . oo . . u."

Passing through the shed, the ORIZABA (sparkling in her sprightliness,) stood before us. A long-arm crane was reaching into the bowels of the ship, and whining tackle shrilled as a basket filled with lumber came to the surface and swung high above the heads of the marching men. A startled command rang out: "Give 'way to the right"!! . . . With a clatter falling lumber slipped from the sling and crashed to the dock . . . "Steady there!" . . . and the line proceeded. A familiar figure walked across the platform and bent over the dead form of one of our men caught under the timber . . . With the man in his arms, he walked away . . . (It was months before I knew he was my brother who had carried away the unfortunate one.)[1]

Grinding wheels whirred; tugs lurched and churned up the water . . . We were moving down the river as the day faded away . . Slashing water lulled me to sleep . . . Then a high sea ran . . . and the Azores slipped gently by. . . . Eight days of fury bore down upon us. Water-swept decks made even the stifling hold preferable . . And the boat sprang a leak.

Day after day the pumps worked away at the seepage. A note of uncertainty marked the tones of the sailors . . . "Let's go down in the hold", I said to Smith . . . "The sailors think the ship is going to sink." . . . (Smith, sea-sick unto death) "O! that's a thing the sailor's will have to worry about . . . I'm not!"

On and on we floated with slashing waves striking at deck-tops. Two sailors struggled across the quarter-deck. A wave lashed at them. One fell with a fractured leg.

But the fury relented, and the sea became quiet. A crowd closed in near to a flag-draped form. A chaplain read, and the form slipped into the sea. Another boy failed to reach home!

I turned away to seek out Smith, who was now retrospective following his illness. The bright sunshine invited us to the deck, and our talk turned from the battlefields of the past to the battlefields of the future. Our habiliments of war yet must be worn, but in spirit we had discarded them for the robes of peace. Dreams supplant recollections. Memories, sacrifices, sufferings are only the ground-work to us for duties and responsibilities yet to come. "Peace hath victories no less renowned than war."

Radiograms posted on the bulletin board announced: "Wilson returning from Peace Conference . . . Will land at Boston" . . . "Parley discusses Fourteen Points" . . . and Smith quoted from Tennyson:—

1 This is the only mention in the book of Christopher Emmett's younger brother, Thomas Addis Emmett, Jr., who also served in France in the A. E. F. Known as "Archie," his brother was mustered in on April 14, 1918 and released from the Service on July 26, 1919 at Camp Travis. U.S. Department of Veterans Affairs BIRLS Death File \ u 0022C 1850 – 2010.

"For I dipt into the future,
far as human eye could see:

Saw the vision of the world and
all the wonders that would be:

Heard the heavens fill with shouting,
and there rained a ghastly dew

From the nation's airy navies,
grappling in the central blue;

Till the war drums throbbed no
longer and the battle flags were furled

In a parliament of man, the federation
of the world."

"Too good to be true!" . . . But had not we fought for the democracy of the world? Had not autocracy given 'way to right? Would not life now be on the plane of comradship and mutual international helpfulness? Had not the one voice of a grateful people altruistically proclaimed: "When they come home, nothing will be too good for them?" Would not the American people hold it a sacred duty to see this goal realized? Had not a generous government already evidenced its sincerity by paying in cash the contemplated profits in industry's canceled war-contracts? . . . but the hazy, smoky outlines of America—home!—came out of the west to greet us.

Tear-stained faces leaned over the rail. A flag-be-decked boat plied close alongside with colors fluttering in profusion. A heraldic banner floated: "Welcome home". Fur-clad women looked upward venturing a feeble wave of the hand.

The ORIZABA docked in New York . . . Down the gangplank . . . into the SUSQUEHANNA, again, after nearly a year . . . and we crossed the river to sleep in Camp Merritt, which rested tranquilly under a blanket of snow.

When morning came the trumpet called again:

"I can't get 'em up,
I can't get 'em up,
I can't get 'em up in the morning:
I can't get 'em up,
I can't get 'em up,
I can't get 'em up at all:

Corporals worse than privates,
Sergeants worse than corporals,
Lieutenants worse than the sergeants,
But the captain's worst of all."

Lethargically answering the bugle's demand, disgusted with the efforts to keep up military appearances, half-dressed soldiers meandered to their first and last formation under a commander only acquired that morning. The day was clear and cold with its carpet of white under foot. The slow 'crunch', 'crunch' of our converging feet intensified the poignancy of the thermometer's fall. Wholly unmilitary, contemptuous in spirit, some leggin-less, shirtless, these soldiers who had once taken the keenest interest in their sprightliness, reported before a company commander who had been left at home and who could not understand the yearnings of returning soldiers to cast aside the mask and be natural once more.

I had come to formation minus blouse and leggings. My shoes were untied. With my right hand in my pocket and cigarette in mouth, I reported my squad: "O! Sarge, about six or seven privates absent." The face of the captain went livid. This was too great a strain on his military dignity . . . "Dismissed! And, sergeant, don't ever form this company of hoodlums again. How they won a war is beyond me." . .

We marched down the station platform the following morning. The red bricks from which the snow had just been removed showed in pointed contrast with the clear clean, sparkling, feathery ice. A car had parked at the station curb. An old man, pale of face, bowed with years, the frost of many winters showing from under his high top-hat, feebly alighted from the car, and with cane tapping on the bricks, came to my side. A far-away look was in his eyes. With a trembling, feeble voice, he asked, half expectantly, half fearfully: "Have you seen my son? . . . They say he is not coming back . . . but I know he will some day . . . I come down each day to see the boys who pass . . . He'll be with them some day."

His chauffeur took him gently but firmly by the arm and led him away . . . "Sir; more boys are coming tomorrow. Perhaps, he'll be with them . . . We'll come back."

As the train glided from the station I raised my hand to him in salute . . . to that spirit . . . to that which was gone . . . And the old gentleman,—every inch a gentleman, who had also suffered the irreparable ravages of war,—with raised top-hat showing a crown of snow, bowed with dignity.

Washington . . . Atlanta . . . New Orleans, and familiar scenes . . . Houston . . . San Antonio . . . And we marched into the streets . . . No one cheered . . . We wondered why! . . . Our file enmeshed with the traffic on the thoroughfare, and we saw that soldiers, uniforms, no longer commanded the enthusiasm of the passers-by. Trucks, automobiles, wagons drawn in commerce, . . .even the city's scavenger-wagons, fetid and reeky, . . . pressed insolently close to our tramping feet as their drivers called to us to give 'way to the right.

We sat down at Camp Travis before the identical barrack where I had spent my first night in the army . . . and of all those of the old gang only Obar and I, now, had returned.

With assignment to a company for discharge, under command of officers who regretted the misfortune of having had to stay at home, they made 'the path out', that 'last long mile', as pleasant and expeditious as the turning of the wheels of army machinery will permit.

The last reveille . . . The roll was called . . . but I was thinking . . . thinking of the year gone by . . . of names, names no longer answering . . . I was startled when I alone answered: "Here" . . . And we marched away . . . to the paymaster and American money . . . to discharge certificates . . . We pinned on red chevrons . . . "Ha . . . ha! We're out of the army now" . . .

My genial friend, the new commander, appeared before me: "Come on back . . . Re-enlist with me!" . . . Raising my hand I thumbed my nose; my fingers wriggled. "*z*!?; you can go to ". . . He looked at my red chevrons and smiled: "I might have known you'd do it!" . . . In the roadway he stood looking after us as we moved toward the exit-gate of Camp Travis.

Trucks crunched the cinder-paths behind us . . . Horns rasped their warnings: "Give 'way to the right!" . . . And we reached the gate . . . "Red chevrons?" . . . "Pass." . . . Outside we hesitated . . . We would take one more view.

Then from out the camp floated the faltering, quavering notes,—a lone bugle! As with the rising tones the vista of the camp before me faded. In its place there came before my eyes faint familiar figures, khaki clad . . . mud spattered . . . helmets askew . . . Rifles fell from their beckoning hands . . . The bugle tones, at first timid, now pitched higher, and they arose as if from the top of the hill . . . these khaki boys, O! so far away!!! . . . Their calls floated across battered fields, attuning for flight against wire-strewn plains, . . . coursing westward over vine-clad dales and castled hills, soaring over expansive seas and echoing down a continent, . . . the voices of Clarence . . . of Tip . . . Molly . . . they whispered to me:

"Farewell dear friend,
I love you so;
That to say 'goodbye' brings
grief no words can tell:
My love is yours for weal or woe,
Dear friend of mine, 'Farewell."

* * *

"Taxi, soldiers? . . . "Let's ride, Ed. I feel too weak to walk."

Fini la guerre.

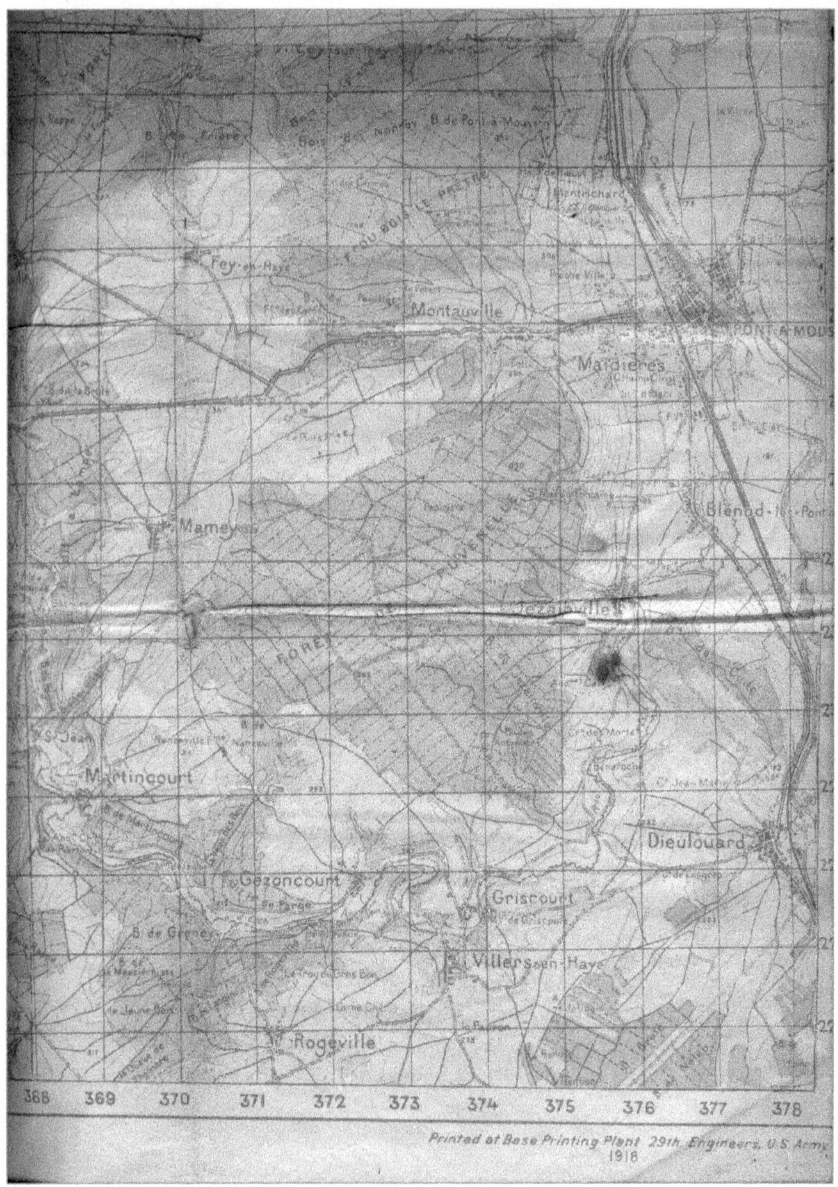

Original combat map carried in the St. Mihiel Offensive. Note the village of Fey-en-Haye in the upper left in the center of the area where Emmett and Company L advanced through the German lines beginning on September 12, 1918. This is a U. S. Army printed and issued map of the area on the eastern half of the AEF attack. Note the mud and sweat stains and the printing unit. Across the map in blue grease pencil are the attack boundaries of the US I Corps and US IV Corps. Thiaucort Map Sheet. 1/50,000. Stieghan Collection.

Editor's Notes

Sergeant Emmett returned to the United States as a casual after spending weeks in a muddy convalescent camp in France recovering from wounds and pneumonia caused by breathing poison gas. He did not march with his Company L comrades following in the dust of the boots of retreating Germans back into their homeland. He missed Occupation duty along the Rhine River and reflecting with his mess mates on their battles over steins of German beer. He sailed westward back across the Atlantic with a boatload of unruly strangers who cared little about military discipline, about each other, and even less about the temporary leaders trying to get them to act like Soldiers.

The troopship that took Emmett back to New York Harbor. USS Orizaba returning troops from Europe to the United States in 1919 with her dazzle paint scheme hastily painted over. U. S. Army Signal Corps photo.

Demobilized with other strangers at the same post outside San Antonio where he was turned into a Soldier eleven months before, Emmett found himself on the street almost invisible to the crowds who had cheered the sight of a brave recruit when he entered those same gates. No parades. No triumphant train ride back with his buddies to East Texas to be greeted by the home folks. His unheralded return would be similar to that of the men discharged from an individual twelve-month rotation from service in Korea or Vietnam. Coming home must have been bittersweet for former Sergeant Christopher Emmett. Imagine his first day returning to his desk as a claims attorney for a railroad. By 1920, he was living as a lodger in the home of his buddy, Peyton Lane, in Jacksonville, Texas. If that name is familiar, Lane was the friend who shared lunch with Emmett on the first page of his story. He finally shook off his bachelorhood by marrying Margaret Craig in 1922.

Doughboy graffiti: a German monument honoring the Prussian Royal Family, the Hohenzollerns, defaced. The T-O symbol at the top was on the divisional patch of the 90th "Texas-Oklahoma" Division, also called the "Tough 'Ombres." Stieghan Collection.

For thirty years, Emmett served in the legal department of the Texas and New Orleans Railway, a division of the Southern and Pacific Railroad, largely from offices in San Antonio or Houston. He spent much of his spare time collecting rare maps and drawings, organizing and leading historical societies, and writing award-winning histories or historical novels. History, particularly Texas History, was his passion. When the United States again entered a world War, Emmett registered for the draft a second time at the age of 56. He continued to represent the railroad throughout the war and until his retirement.[2]

2 (Ed.) Upon his registration interview, Emmett asserted that he had no middle name, therefore no middle initial. Christopher Emmett. Registration Card. Form 1. Serial Number U735. D. S. S. Form 1.

Following retirement from the Southern Pacific Railroad, Emmett moved to Sante Fe, New Mexico, for a number of years and became interested in the history of that state. His last published work is a history of Fort Union, New Mexico Territory. He returned to Texas late in life and died in Dallas, October 19, 1971. He is buried in Oakwood Cemetery, Hamilton, Texas.[3]

Recall his flashback in the introduction? He was proud to be included among those in rumpled wool uniforms who marched and sang together in parades celebrating their reunions and service. It didn't matter whether they knew each other in the Army. All that mattered is that they all served so they were by extension veterans of the same experience. They all wore the uniform and served as a Soldier. A bystander's cold reaction to the pride and swagger of these men stepping together down the street did something to Emmett. Nearly an entire generation had grown into manhood without understanding the sacrifices of the youth of his age. He realized that he had a story to tell.

Emmett had obviously kept notes or a diary during his service. As stated in his Forward, he bagan the first version of his reminiscences a week after his discharge while his memories were fresh. Many years later, he decided to dust off his wartime notes and type an unvarnished account of his participation in the Great War. Why did he wait fifteen years to write *Give Way to the* Right? He had already published the book *Texas Camel Tales* and was a noted historian around the state. He did not consider himself famous or a hero, but he was proud of the way that he had served and realized that he missed his buddies. Some survived and too many didn't. He was wounded and gassed and had been chosen to lead his peers by those who recognized his maturity and natural leadership, in spite of his often-sour attitude.

His story of service is not that of a hero. He "ran the blockade" to get outside of camps on two continents to get a hot meal, a drink, or just to see the sights. With the exception of one official pass from his company commander, which wasn't accepted by a guard, he often led small groups to

3 (Ed.) Emmett, Christopher, *Texas History Online.*

sneak out of camp to sightsee in New York City and numerous towns and villages in England and France. Once he determined he would not serve as a pilot or intelligence officer, he decided to become another private and get by with his comrades. Occasionally, though, he could not help himself. He was an educated man and a professional in civilian life. He began his adventure in uniform at thirty-two years of age and played a young man's game. He enjoyed the new sights of his own country, England, and France, and had an appreciation of the history and culture from his college days. In his book he quotes from classic literature and poetry, perhaps created prose of his own, and also shared the ribald verses of his fellow Soldiers' songs with his readers. Emmett wrote a tale of roughing it in the Army, added spice with Soldier's songs and classic poetry, and in the process created a classic tale of the Doughboy experience.[4]

On several occasions, Emmett mentions his "automatic." Shown is an original U. S. Pistol Model 1911 made by Colt. The magazine contains seven rounds of potent .45" A. C. P. ammunition. With a few changes, this pistol was the principal sidearm of the United States military from 192 until after Operation Desert Storm in 1991. Stieghan Collection.

4 (Ed.) Emmett, Christopher. Certificate of Death. State of Texas. State File No. 70377. Texas Department of Health, Bureau of Vital Statistics.

Weapons and equipment used by Sergeant Christopher Emmett and his fellow Infantrymen of the 90th Division in World War I. Described from left to right, from top to bottom. Underneath are a raincoat or "slicker," a wool overcoat, and a wool blanket. The rifle is a U. S. Rifle Model 1917 made at Remington, the pistol is a US Pistol Model 1911 made at Colt's Patent Fire Arms Manufacturing Company, and a French Model 1915 C. S. R. G. Chauchat Automatic Rifle with a loaded spare magazine on the blanket. Also visible are a trench periscope, haversacks, gas masks, "Pershing Boots," wool wrap puttees, hand grenades, rifle grenades and a launcher, a helmet with "corned Willie" and "Monkey Meat," a canteen with web carrier, a Model 1918 Trench Knife, a cartridge belt, and a first aid pouch. Stieghan Collection.

Roster of L Company, 359th Infantry

Company Roster

ALBRECHT. GEORGE J., Private, Princeton, Millelacs, Minn.

ALLEN, JAMES O., Sergeant, Arp, Smith County, Texas.

ALLEN, DOCTOR T., Private, General Delivery, Tyler, Smith County, Texas.

ALLEY, ROSS H., Private, Killed in action. Address of Dependent, Leesville, Gonzales, Texas.

ALLISON, SAM, Private, Route No.1, Biardstown, Lamar County, Texas.

ALMY, ORLIN W., Private, 1121 West Agarita, San Antonio, Texas.

ALSTON, LONNIE G., Private, P. O. Box 42, Canutillo, El Paso, Texas.

ANCIAUX, CHARLES, Private, 1039 Keokuk St., Iowa City, Iowa.

ANDERSON, WILLIAM C., Private, Lubbock, Texas.

ARRANT, ALFRED, Private, Kerrville, Kerr County, Texas.

BAKER, JIM P., Private, Silverton, Texas.

BAILEY, JAMES A., Private, Ponta, Texas.

BARRETT, BEN T., Corporal, Killed in action. Sept. 12, 19IS. Address of Dependent, Henderson, Rusk, Texas.

BARRON, LEONARD C., Corporal, 8 Fisk Apts., 328 W. Missouri St., El Paso, Texas.

BARRON, WILLIAM, Corporal, 1221 S. College St., Tyler, Smith County, Texas.

BARTLETT, EDWARD R., Private, Bartlette, Bell County, Texas.

BARTON, BEN, Private, R. F. D. No. 2, Abbott Hill, Texas.

BAUER, JOSEPH A., Private. Died of wounds. Sept. 13. MM. 1210 N. 6th St., Nankato, Blue Earth County, Minn.

BAUER, RUDOLPH C., Private, Stockholm, Pepin, Wisconsin.

BEALL, OLLIE, Private, Longview, Gregg, Texas.

BEASLEY, THOMAS DANIEL, Private, Lake Creek, Delta County, Texas.

BENEDICT, CHARLES E., Corporal, 505 West 25th St., Ft. Worth. Texas.

BENGE, HEIMAN S., Corporal,Tivoli, Texas.

BETZ, LOUIS F., Private, 1381—E. 78th St., Los Angeles, Calif.

BOLING, FRANK MIRON, Private, Died in Service, Oct. 20, 19IS, Address of Dependent. 417 East Gray St., Norman, Okla.

BOUNDS, ROBERT B., Private, 405 Lake Shore Drive, Port Arthur, Texas.

BOWLES, WILLIE H., Private, R.F.D. No. 3, Lindale, Texas.

BRASWELL. MONROE P., Private, DeKolb, Bowie County, Texas.

BRODKEY, CHARLES E., Private, Chiropractic Psychopathic Sanatorium, Davenport, Iowa.

BROWN, LOUIS EUAS, Private, 707 E. Euclid Ave., San Antonio, Texas.

BROWN, LUKE L., Private, Eakly, Caddo County, Oklahoma.

BRUCE, CLAYBORNE ANDERSON, Private, Midland, Texas.

BRUCH, WM. A., Private, Route No. 1, Montevideo, Chippewa County, Minn.

BRUNING, WM. A., Private 413 Greggs Ave., Grand Forks, North Dakota.

BURNS, OMERA D., Private, Troop, Smith County, Texas.

BUSLER, JAY, Private, Deer River, Minn.

CAMPBELL, GEORGE CALVERT, Private, 3121 Eighth St., Port Arthur, Texas.

CARROLL. JAMES M., Private, Tyler, Smith County, Texas.

CASTILLO, GERMAN, Private, 1011 S. Adams St., San Antonio, Texas.

CHAMBERS, JAMES LESTER, Private, 1005 E. High St., Terrell, Kaufman County, Texas.

CHAPMAN, EARL Z., Private. 207 Fallen Ave., Charles City, Iowa.

CHLOUPEK, RUDOLPH, Private. (Died of wounds received in action Oct. 30, 1918). Gary, Norman County, Minn.

CLOUD. CHARLES F., Private, Brady, McCulloch County, Texas.

COATS. NORVELL WADE, Corporal, Mt. Enterprise, Rusk County, Texas.

COLE, SAM J., Pvt. 1st Class, Turney, Cherokee County, Texas.

COMSTOCK. EMERY E., Private Route 3, Foley, Morrison County, Minn.

CONWAY. VIRGLE G., Bugler, Elderville, Gregg County, Texas.

COOPER. Gustave C., Private, Window, Texas.

CROW, CHAS. JAMES, Private P.O. Box 315, Rusk, Cherokee County, Texas.

CURRY, FRANKLIN P., Corporal, Lindale, Smith County, Texas.

DANNER. ORLEY B., Private, Rural Route No. 2, Astoria, Fulton County, Ill.

DAVES, JAMES H., Mess Sgt., 2108 Grape St., Abilene, Texas.

DEASON, OTIS TERRELL, Corporal, 310 E. Arsenal St., San Antonio, Texas.

DICKERSON. WM. B., Private, Gladwater, Gregg County, Texas.

DIETZE. HERMAN LEROY, Mechanic, 219 Roberts Blvd., Univ. Park Addition, Dallas, Texas.

DRISCOLL, LEO P., Private, Bernard, Jackson County, Iowa.

DUNN, KENNETH HAYS, Private, Died after discharge at Victoria, Texas. 899 S. Bridge St., Victoria, Texas.

DURHAM, HILL CUNTON, Woodville, Texas.

DURHAM, JIM E., Cook. Box U., Breckenridgc, Stephens, Texas.

EMMETT, CHRISTOPHER, Private, 315 Hicks Bldg., San Antonio, Texas.

ERVIN, ELBERT THEODORE, Private, Sumner, Lamar County, Texas.

EVANS, WILL, Private, Alto, Cherokee County, Texas.

FELTON, CLAY, Private, Weatherford, Custer County, Okla.

FERGUSON, HENRY M., Private, 1001 Praetorian Bldg., Dallas, Texas.

FERRELL, ROSCOE D., Private 1st Class, R.F.D. No. 4, Lindale, Smith County, Texas.

FLESNER, RIKUS II., Private, Killed in action Sept. 12, Mis. 139 No. 12th St., Quincy, Ill.

FLORES, GREGORIO, Private, 2612 San Dario Ave., Laredo, Texas.

FLY, MONTGOMERY, Lt. Killed in action.

FLYNT, ROBERT O., Sgt., Flint, Smith County, Texas.

FOWLER, EUGENE, Corporal, 214 Spruce St., Texarkana, Texas.

FUCHS, PETER J., Corporal, Underwood, McLean County, North Dakota.

GARNER, ARGIE E., Supply Sgt., 520 S. Ragsdale, Jacksonville, Texas.

GENEREAU, JOHN T., Private, 468 East First St., Duluth, Minn.

GILL, JUSTIN SIDNEY, Private, Airdale, Bosque County, Texas.

GLIDDEN, ROBERT J., Private, Rusk, Cherokee County, Texas.

GOBER, GEORGE A., Corporal 214 B. Bolton Street, Jacksonville, Texas.

GOENS, ERNEST S., Sgt., 316 W. Erwin, Tyler, Texas.

GOFF, GEORGE H., Private, Alto, Texas.

GOODSON, WM. L. (BUI), Private, 2312 Ethel Ave., Waco, Texas.

GOULD, JOHN EDWIN, Private, Teneha, Shelby County, Texas.

GOWIN, GEORGE E., Private, Jacksonville, Cherokee County, Texas.

GRAY, CHARLES A., Private, Jackson, Minnesota.

GREEN, HARRY J., Private, 317 Sherman Ave., Council Bluffs, Iowa.

GREEN, ROBERT KLINE, Private, Mt. Enterprise and Rusk, Texas.

GRESHAM, JESSE M., Corporal, Killed in action, Sept. 12 1918. K.K.I J. No. 6, Conway, Faulkner County, Arkansas.

GROVIER, ARTHUR, Private, Ogden Hotel, Council Bluffs, Iowa.

GROW, HARDY LEWIS . . . .

HABY, ELMER HENRY, Private, Dunlay, Medina County, Texas.

HAMILTON, ROBERT . . . .

HEDLUND, CARL . . . .

HENDERSON, ARCH ALLEN . . . .

HAYGOOD, THOS . . . .

HERRING, BOB . . . .

HEATH, JOHN R . . . .

HILLIN, JOE A. . . . .

HOLCOMB, JAS. KIRBY . . . .

HOWELL, JASPER H. . . .

HIRT, RUDOLPH . . . .

HOUSE, JOHN R., Lieutenant, Box 1452, Lincoln, Nebraska.

HURMANS, JOHN W. . . . .

INGLEY, WILLIE C., Private. Stephensville, Texas.

JACKSON, LAWRENCE W., Corporal, Henderson, Texas.

TAMES, JOHN A., Private, Claudell, New Mexico.

JERNIGAN, DOLPHUS W., Private, Mt. Selman, Texas.

JONES, ALBERT, Private, Jacksonville, Texas.

JONES, IMPSON, Private, Tushka, Oklahoma.

JONES, PETE, Private, Rusk, Cherokee County, Texas.

JONES, WILLIE JOE, Private, 801 North Harwood St., Dallas, Texas.

JOHNSON, FRANCIS V., Private, Wausa, Nebraska.

KALB, AARON, Private, 1512— 14th St., Galveston, Texas.

KENDALL, COLUMBUS L., Private, Warren, Bradley County, Arkansas.

KENRICK, MARVIN. R., Corporal, 4110 Wycliff Ave., Dallas, Texas.

KING, CHARLES FRANCIS, Private, 536 W. 7th St., Dallas, Texas.

KING, EARL PRIDDY, Private, Batesville, Zavala County, Texas.

KING, LLOYD W., Private, P. O. 58, Kaufman, Terrell County, Texas.

KINNE, NOMAN T., Private, 3114 Baker Ave., Bryan, Texas.

KIRKENDALL, JESSE, Private, Sulphur Bluffs, Hopkins County, Texas.

KNUTSEN, MARENUS J., Private, 1320 S. 9th St., Council Bluffs, Iowa.

KNUTSON, CLARENCE A., Private, Pelican Rapids, Otter Tail County, Minn.

KOONCE, KENNETH NEWELL, Private, Rural Route No. 1, Mt. Enterprise, Rusk County, Texas.

KORFIATIS, JOHN T., Private, Died of wounds received in action. Residence at enlistment. Glasgow, Montana.

KRAMER, FRED, Private, 3505-5th Ave., Sioux City, Iowa.

LANDRY, WILLIAM A., Private, 819—15th Street, Modesto, California.

LARSON, CLARENCE L., Private, 3132 Cedar Ave., Minneapolis, Minn.

LARSON, LLOYD L., Private, Decorah, Winneshiek County, Iowa.

LILLEY, JOHN O., Mechanic, Whitehouse, Smith County, Texas.

LLOYD, JOE G., Coporal, Killed in action Nov. 1, 1918. (Henderson, Rusk County, Texas).

LOEB, ALVIN MITCHELL, Private, 3007 Parks Row, Dallas, Texas.

LONBERG, ALBERT O., Private, 445—1st Ave., N.W., Minot, North Dakota.

LOVE, VERNE L., Private, 3526 Page Blvd., St. Louis, Mo.

LUNA, BARTOLO L., Private, Route No. 1, Box 24, Jourdanton, Texas.

McBRIDE, DOUGLAS W., Private, Tyler, Smith County, Texas.

McCAULEY, LON M., Private, Lipan, Hood County, Texas.

McCRARY, BENJAMIN F., Private, Arp, Smith County, Texas.

McDONALD, FRED DENSON, Private, Route No. 1, Mt. Alba, Texas.

McDONALD, LUMA E., Private, P. O. Box 24, Jacksonville, Texas.

McGUIRE. CARL M., Private, 106 South Oak St., Pratt, Kansas.

McHANEY, ROBERT NEWEL, Private, Portland, Oregon.

McKILBEN, CHESTER A., Pivate, Route No. 2, Grand Salme, Van Zandt County, Texas.

McKINNEY, STONEWALL J., Corporal, Jacksonville, Cherokee County, Texas.

McVEY, ROBERT HOMER, Private, 5643 Goodwin Ave., Dallas, Texas.

MATTOX, EMMETT E., Private, Sallisaw, Oklahoma.

MEEKS, DENNIS E., 1st Sgt., 701 Avenue "B", San Antonio, Texas.

MERK, ROBERT MERRIT, Private, 206 S. Broadway, Tyler, Texas.

MESSER, WM. THOMAS, Private, O'Brien, Haskell County, Texas.

METZGER, ELMER W., Private, 5612 Columbia Ave., Dallas, Texas.

MOORE, HENRY, Corporal, Route No. 8, Tyler, Smith County, Texas.

MOORE, JOE HUBERT, Private, P. O. Box 74, Alto, Texas.

MORALES, IGNATIUS, Private, 1109 W. Houston St., San Antonio, Texas.

MORES, SAMUEL G., Private, 2408—5th Ave., Molene, Illinois.

MORRIS, ARTHUR, Private, Route No. 7, Canton. Van Zandt County, Texas.

MORROW, WILLIAM C., Sgt., Killed in action. Manton, Martin County, Texas.

MULKEY, FANK B., Private, Apache, Caddo County, Okla.

MULLER, CHARLES L., Corporal, 202 East Jackson St., Brislow, Okla.

MURROW, CLAY D., Private, 719 First St., Alva, Woods County, Okla.

MYRICK, WM. JENNINGS, Private, Dallas, Texas.

NASH, SYLVESTER J., Sgt., 2034 ½ Texas Ave., Shreveport, La.

NELSON, ERNEST A., Private, 420 N. 37th St., Council Bluffs, Iowa.

NEWCOMB, HARRY, Private, 2521 Clay St., Denver, Colo.

NORMAN, RICHARD C., Cook, Box 266, Corsicana, Texas.

NYSEWANDER, VICTOR H., 1st Lieut., Jonesville, Bartholomew County, Indiana.

O'BAR, JOHN ROBERT, Private, Route 4, Griffin, Texas.

OCHS, BURLON PAUL, Private, Killed in action. Sept. 21, IMS. Address of Dependent, 32 Spruce St., Texarkana, Texas.

O'DONDON, WOLLIA, Private, 805 Seventh St., Des Moines, Iowa.

ODOM, GILBERT, Private, Route 3, Rusk, Texas.

ODOM, ROY, Private, 727 Hill Crest, Dallas, Texas.

OLIVER, CLARENCE EUGENE, Private. Killed in action, Sept. 12, 1918. Address of Dependent. Mt. Enterprise, Rusk County, Texas.

PARIS, JOHN O., Private, 615 N. Ragsdale St., Jacksonville, Texas.

PARKER, CLARENCE, Private, Arp, Texas.

PATTON, JULIAN SMITH, Private, Killed in action, Nov. 3, 1918. Address of dependent. De Sota Hotel, St. Petersburg, Fla.

PEAVEY, NEEDHAM, Private, Mabank, Kaufman County, Texas.

PERRY, WM. J., Private. Killed in action, Sept. 12, IMS. Address of dependent. Jacksonville, Texas.

POOL, ODIE LUCIOUS, Private, Overton, Rusk, Texas.

POOLE, ARTHUR D., Cook, Frankston, Anderson County, Texas.

PRIOR, OLIVER ISAAC, Private, General Delivery, Hallsville, Texas.

PUGH, CLAUD, Private, Hawkins, Wood County, Texas.

RADL, ANDY, Private. Killed in action, Sept. 15, 1918. Sleepy Eye, Minn.

RANKIN, CHARLES P., Mechanic. Killed in action, Sept. 28, 1918. Dependent. Henderson, Rusk County, Texas.

REISDORPH. WARD W., 1411 Hamilton St., Sioux City, Iowa.

REYNOLDS. CARLOS EVANDER, Private, Demmitt, Castro County, Texas.

RIGGINS, JAMES MONROE, Private, Saginaw, Tarrant County, Texas.

ROBERTS, CLYDE W., Private, Big Sandy, Upshur County, Texas.

ROBSHAW, GUY M., Private, Council Bluffs, Iowa.

ROLPH, FRED, Private, Eldora, Hardin County, Iowa.

ROSENBAUM, DAVID, Private, 1517 Forest Ave., Des Moines, Iowa.

ROSS, GEORGE A., Corporal, Rusk, Texas.

SADLER, EARL R., Private. Gatesville, Texas.

SAGE, EDMOND, Private, 902 W. Market St., Normal, Ill.

SANDERS, ALLEN FRANCIS, Private, Route No. 5, Tyler, Texas.

SANDERS, EDWIN E., Private, Wells, Texas.

SANDERS, JAMES T., Private, 109 N. Taylor St., Ft. Worth, Texas.

SANDERS, JULIUS MITCHELL, Private, Route No. 3, Tyler, Texas.

SAVAGE, CALVIN C., Private, Winfried, Texas.

SCHEEPER, HENRY, Private, Tipton, Cedar County, Iowa.

SCOTT, HARRY HOGAN, Private, 839 N. Beckley Ave., Dallas, Texas.

SELF, GEORGE LAVADA, Private, 907 N. Church St., McKinney, Texas.

SELLERS, OLIVER SHEID, Private, c/o Arthur G. McJuan, 751—B. S. Figuera, Los Angeles, California.

SHARMAN, GORDON O., Private, Sudan, Lamb County, Texas.

SHARMAN, THOS. S., Private, Route No. 1, Lindale, Texas.

SINGLETON, GEORGE R., Sgt., 522 Cherry St., Jacksonville, Texas.

SLOAN, JOHN MAX, Private, 2221 Maple St., Little Rock, Arkansas.

SMITH, JOHN DAVID, Private, 416 W. Harris St., San Angelo, Texas.

SPEED, ARTHUR ERNEST, Private, Killed in action, Sept. 15, 1918. Rosser, Kaufman County, Texas.

SPIVEY, KIRBY, Private, Died in service, Dependent, Rusk, Texas.

STEENERSON, ROBERT, Private, 619 First Ave., Chisholm, Minn.

STERLING, BENJAMIN F., Private, 1903 Portsmouth, Houston, Texas.

STOCK, FRANK C., Private, St. Paul, Minn.

STOCKTON, CHESTER E., Private, R.F.D. No. 5, Box 72, Troup, Texas.

STONE, TED S., Private, Ainsworth, Iowa.

STRUNK, ERIC STANLEY, Private, Oakland, Texas.

SUMNER, CHARLES N., Private, c/o Taylor Dentists, Waterloow, Iowa.

SURATT, JAMES HARVEY, Private, Route 6, Box 439, Ft. Worth, Texas.

TASA, OLE O., Private, c/o N. J. Nelson, 503 W. Sharp Ave., Spokane, Wash.

TAYLOR, BEN C., Cook, R.R. No. 9, Trier, Texas.

TEBBE, GEORGE., Private. Killed in action, Sept. 12, 1918. 1815 McKinney Ave., Dallas, Texas.

TELKAMP, CARL, Private, 4243 Tuttle Ave., Dallas, Texas.

THOMAS, CHARLES W., Private, 325 W. Second St., Tyler, Texas.

THOMES, HARVEY A., Private, Kingston, Des Moines, Iowa.

THOMPSON, LUTHER W., Sgt., 413 West Elm St., Tyler, Texas.

THOMPSON, ONA W., Private, Gallatin, Cherokee County, Texas.

THOMPSON, WM. B., Sgt., Rusk, Texas.

TUCKER, WM. P., Sgt., 2738 Ave. "F", Ft. Worth, Texas.

VOGT, THEODORE, Private, R.F.D. No. 4, Carroll, Iowa.

WALKER, GEORGE E., Private, Kilgore, Texas.

WALKER, PRINCE, Private, R.F.D. No. 4, Kaufman, Texas.

WALLACE, WILLIE, Private, Cushing, Texas.

WALSH, GORDON, Private 1st Class, R.F.D. No. 9, Box 60, Tyler, Texas.

WARD, JAMES E., Bugler, P. O. Box 417, Jacksonville, Texas.

WEATHERFORD, CHARLIE C., Private, Box 36, Devers, Liberty County, Texas.

WEBER, FERDINAND HENRY, Private, Marfa, Texas.

WELLS, DOCK D., Private, R.F.D. No. 1, Garden Valley, Texas.

WELLS, ERNEST C., Corporal, Winona, Texas.

WHITAKER, BERRY M., Captain, 914 West 26th St., Austin, Texas.

WHITE, HENRY O., Private, San Saba, Texas.

WILKES, RAY, Private, Nitre Home, Baxter, Arkansas.

WINROD, THOMAS A., Private, Box 543, Parsons, Labette County, Kansas.

WOLFE, CHAS. H., Private, 526 S. First St., East Cedar Rapids, Iowa.

WOODRUFF, VIRDIE, C., Private, P. O. Box 1035, 3206 Buchanan St., Wichita Falls, Texas.

WORRELL, WM. OSCAR, Private, Henderson, Texas.

WYLIE, GEORGE CURTIS, Private. Pine Hill, Texas.

WYLIE, RUSSELL EWING, Private, Killed in action, Nov. 11, 1918. Henderson, Texas.

YOUNG, WINFORD GRADY, Private, R.F.D. No. 1, Henderson, Texas.

Bibliography

Allen, Hervey. *Toward the Flame*. New York: George H. Doran Co, 1926.

American Battlefield Monuments Commission. *Amrican Armies and Battlefields in Europe*. Washington, D. C.: Government Printing Office, 1938.

Army War College. *Manual for Commanders of Infantry Platoons. Translated from the French (Edition of 1917) at the Army War College.* Washington, D. C.: Government Printing Office, July 6, 1917

Bernasconi, Commander Jeffery J. "Cannon Fodder or Corps d' Elite: The American Expeditionary Force in the Great War." Pickle Publishing Partners, 2014. Accessed 3 June 2017.

Bertke, Donald A. Gordon Smith, Don Kindell, *World War II Sea War, Vol. 7: The Allies Strike Back*, Dayton, Ohio: Bertke Publications, 2014.

Burdick, Major Henry H., 318th Infantry [Regiment, commanding 1st Battalion], A. E. F. 1919. "Development of the Half-Platoon as an Elementary Unit." *Infantry Journal* XV, no. 10 (April): 799-809.

Chastaine, Captain Ben H. *Story of the 36th: The Experiences of the 36th Division in the World War.* Harlow Publishing Company. Oklahoma City, OK, 1920.

Cullen, Lieutenant Commander Glen T. "Preparing for Battle: Learning Lessons in the U. S. Army During World War I." Pickle Publishing Partners, 2014. Accessed 3 June 2017.

Emmett, Chris. *Texas Camel Tales* (1932),

____________. *Give 'Way to the Right.* San Antonio, Texas: The Naylor Company, 1934.

____________. *Texas As It Was Then: A Conspectus of the Adventures of Cabeza de Vaca and La Salle in Texas.* San Antonio, Texas: The Naylor Company, 1935 and 1936.

____________. *The General and the Poet.* San Antonia, Texas: The Naylor Company, 1937.

____________. *Shanghai Pierce: A Fair Likeness. The story of Abel Head (Shanghai) Pierce who left Rhode Island penniless and became one of the Big Pasture Men of southern Texas.* Norman, Oklahoma: The University of Oklahoma Press, 1953.

____________. *In the Path of Events.* Waco, Texas: Jones & Morrison, 1959.

____________. *Fort Union and the Winning of the Southwest.* Norman, Oklahoma, 1965.Emmett, Christopher, *Texas History Online.*

General Headquarters, American Expeditionary Force. *Instructions for the Offensive Combat of Small Units.* A copy of the April 1918 training circular, signed by General James G. Harbord, Chief of Staff of the A. E. F., and approved by General John J. Pershing, Commander of the A. E. F., April 1918.

____________________, _________________________. *Instructions for the Offensive Combat of Small Units, Prepared at the General Headquarters, American Expeditionary Forces, France, from an official French Document of January 2, 1918.* A[jutant] G[eneral]. Printing Department [Paris], G. H. Q. A. E. F. [General Headquarters, American Expeditionary Forces], 1918. A copy of the March 1918 training circular, signed by General James G. Harbord, Chief of Staff of the A. E. F., and approved by General John J. Pershing, Commander of the A. E. F., was provided to the author by a private collector in 1995. The document is hand typed and the formation illustrations are hand drawn.

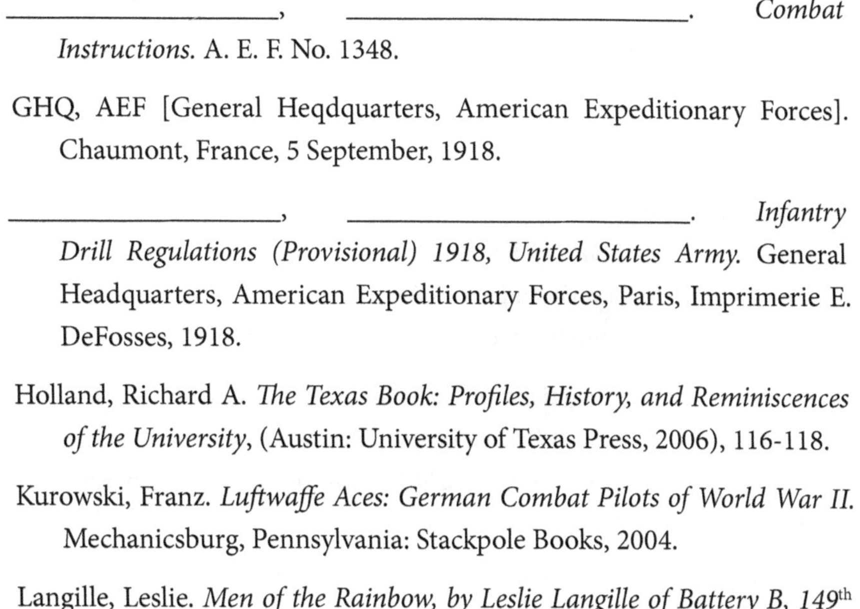

———, ———. *Combat Instructions.* A. E. F. No. 1348.

GHQ, AEF [General Heqdquarters, American Expeditionary Forces]. Chaumont, France, 5 September, 1918.

———, ———. *Infantry Drill Regulations (Provisional) 1918, United States Army.* General Headquarters, American Expeditionary Forces, Paris, Imprimerie E. DeFosses, 1918.

Holland, Richard A. *The Texas Book: Profiles, History, and Reminiscences of the University*, (Austin: University of Texas Press, 2006), 116-118.

Kurowski, Franz. *Luftwaffe Aces: German Combat Pilots of World War II.* Mechanicsburg, Pennsylvania: Stackpole Books, 2004.

Langille, Leslie. *Men of the Rainbow, by Leslie Langille of Battery B, 149th Field Artillery, U. S. A. 42nd (Rainbow) Division, A. E. F.*, Chicago, Illinois: The O'Sullivan Publishing House, 1934.

Lawrence , Joseph Douglas. *Fighting Soldier: The AEF in 1918.* Edited by Robert H. Ferrell. Boulder: Colorado Associated University Press, 1985.

O'Conner, Richard. *Black Jack Pershing.* Garden City, New York: Doubleday and Company, Inc., 1961.

Pershing, General John J. "General Pershing's Official Story; Battles Fought by the American Armies in France from their Organization until the Fall of Sedan." *Infantry Journal* XV, no. 9 (March): 692.

Pershing, John J. *My Experiences in the World War.* 2 vols. New York: Frederick A. Stokes,1931.

Stallings, Laurence. *Plumes.* New York: Harcourt, Brace and Company, 1924.

Stieghan, David S., ed. *Over the Top!* Dahlonega, Georgia: University of North Georgia Press, 2017.

Strickland, Riley. *Adventures of the A.E.F. Soldier.* Austin, Texas: Von-Boeckmann-Jones Co., 1920.

Texas History Online, accessed March 3, 2014. https://tshaonline.org/handbook/online/articles/fem02

United States Army in the World War, 1917-1919. Organization of the American Expeditionary Forces. Volume 1. Washington, D. C.: Center of Military History, United States Army, 1988. http://www.theshipslist.com/ships/lines/bluefunnel.shtml Accessed 9 March 2017.

War Department. *Instructions on the Offensive Conduct of Small Units, Translated from French Edition of 1916, Edited at the Army War College.* War Department Document No. 583. Washington, D. C.: Government Printing Office, May 1917.

War Department. *Instructions for the Offensive Combat of Small Units, Prepared at the General Headquarters, American Expeditionary Forces, France, from an official French Document of January 2, 1918.* War Department, War Plans Division. Document No. 802, Office of the Adjutant General, May, 1918.

War Department. *Combat Instructions*, A. E. F. No. 1348, War Plans Division, October 1918. Also see: War Department Document No. 868, Office of the Adjutant General. War Department, Washington, October 5, 1918.

Wythe, Major George. *A History of the 90th Division.* San Antonio, Texas: The 90th Division Association, 1920.

Official Records

1900 United States Census for Christopher Emmett. Comanche County, Texas.

1910 United States Census for Christopher Emmett. Hamilton City, Hamilton County, Texas.

1920 United States Census for Christopher Emmett. Jacksonville, Cherokee County, Texas.

1940 United States Census for Christopher Emmett. Olmos Park City, Bexar County, Texas.

Christopher Emmett. University of Texas. General register of the students and former students of the University of Texas, 1917 / compiled for the Ex-student's Association by W.J. Maxwell under the general direction of John a. Lomax, Secretary. Ancestry.com. *U.S., School Yearbooks, 1880- 2012.* Provo, UT, USA: Ancestry.com Operations, Inc., 2010.

Christopher Emmett. Draft Card. 1917. Texas. Cherokee County. Draft Card E. Registration State: *Texas*; Registration County: *Cherokee*; Roll: *1952402*

Christopher Emmett. Draft Card. 1942. Texas. Chastain, Claude Alexander - Galleher, Neudeget Craven. The National Archives at St. Louis; St. Louis, Missouri; *Draft Registration Cards for Fourth Registration for Texas, 04/27/1942 - 04/27/1942*; NAI Number: *576252*; Record Group Title: *Records of the Selective Service System*; Record Group Number: *14.*

Christopher Emmett. Texas. Death Certificates, 1903-1982 for Dallas. 1917. Oct-Dec. Accessed on Ancestry.com. *Texas, Death Certificates, 1903-1982.* Provo, UT, USA: Ancestry.com Operations, Inc., 2013.

www.ingramcontent.com/pod-product-compliance
Lightning Source LLC
Chambersburg PA
CBHW021550210326
41599CB00010B/386

* 9 7 8 1 9 4 0 7 7 1 4 4 1 *